MW01518513

NATO/CCMS Workshop on
 Oil Spill Response
Oil spill response : a
 global perspective
       c2008.

# Oil Spill Response: A Global Perspective

# NATO Science for Peace and Security Series

This Series presents the results of scientific meetings supported under the NATO Programme: Science for Peace and Security (SPS).

The NATO SPS Programme supports meetings in the following Key Priority areas: (1) Defence Against Terrorism; (2) Countering other Threats to Security and (3) NATO, Partner and Mediterranean Dialogue Country Priorities. The types of meeting supported are generally "Advanced Study Institutes" and "Advanced Research Workshops". The NATO SPS Series collects together the results of these meetings. The meetings are co-organized by scientists from NATO countries and scientists from NATO's "Partner" or "Mediterranean Dialogue" countries. The observations and recommendations made at the meetings, as well as the contents of the volumes in the Series, reflect those of parti-cipants and contributors only; they should not necessarily be regarded as reflecting NATO views or policy.

**Advanced Study Institutes (ASI)** are high-level tutorial courses intended to convey the latest developments in a subject to an advanced-level audience

**Advanced Research Workshops (ARW)** are expert meetings where an intense but informal exchange of views at the frontiers of a subject aims at identifying directions for future action

Following a transformation of the programme in 2006 the Series has been re-named and re-organised. Recent volumes on topics not related to security, which result from meetings supported under the programme earlier, may be found in the NATO Science Series.

The Series is published by IOS Press, Amsterdam, and Springer, Dordrecht, in conjunction with the NATO Public Diplomacy Division.

### Sub-Series

| | | |
|---|---|---|
| A. | Chemistry and Biology | Springer |
| B. | Physics and Biophysics | Springer |
| C. | Environmental Security | Springer |
| D. | Information and Communication Security | IOS Press |
| E. | Human and Societal Dynamics | IOS Press |

http://www.nato.int/science
http://www.springer.com
http://www.iospress.nl

**Series C: Environmental Security**

# Oil Spill Response: A Global Perspective

edited by

## W.F. Davidson
NATO/CCMS National Representative,
Director (National Facilities) National Research Council,
Ottawa, Ontario, Canada

## K. Lee
Centre for Offshore Oil and Gas Environmental Research,
Fisheries and Oceans Canada,
Dartmouth, Nova Scotia, Canada

and

## A. Cogswell
Centre for Offshore Oil and Gas Environmental Research,
Fisheries and Oceans Canada,
Dartmouth, Nova Scotia, Canada

 Springer

Published in cooperation with NATO Public Diplomacy Division

Proceedings of the NATO CCMS Workshop on
Oil Spill Response
Dartmouth, Nova Scotia, Canada
11–13 October 2006


Library of Congress Control Number: 2008928526




ISBN 978-1-4020-8564-2 (PB)
ISBN 978-1-4020-8563-5 (HB)
ISBN 978-1-4020-8565-9 (e-book)

---

Published by Springer,
P.O. Box 17, 3300 AA Dordrecht, The Netherlands.

*www.springer.com*




*Printed on acid-free paper*

---

# TABLE OF CONTENTS

## SUMMARY REPORT

I would like to welcome you to this Workshop, the third in a series, on oil spill response. My home base is the National Research Council of Canada, Ottawa, and I am co-chair of the Workshop. I am currently the national representative to the NATO Committee on the Challenges of Modern Society (CCMS). The CCMS, which dates from 1969, comes under the Public Diplomacy Division of NATO.

This Workshop takes place under the aegis of NATO's relationship with the Russian Federation: the NATO Russia Council (NRC). As well, there are some 110 participants in attendance, a great testimony to the interest in and concern surrounding the Workshop topic of Oil Spill Response. Seventeen nations are represented: Austria, Belgium, Canada, Denmark, France, Kyrgyz Republic, Latvia, Norway, Poland, Russian Federation, Slovenia, Spain, Sweden, Turkey, United Kingdom, USA and the Former Yugoslav Republic of Macedonia.

I would like to welcome you to Canada and wish you an enjoyable and fruitful stay.

This Workshop is sponsored by NATO, and we have also secured support for this Workshop from the Department of Foreign Affairs and International Trade, the Department of Fisheries and Oceans, and the National Research Council of Canada, for which we are very grateful.

The first workshop took place under the NATO-Russia Council (NRC) Committee on the Challenges of Modern Society in *Horten, Norway in April 2004*. It was attended by 50 participants from Russian Federation, USA, France, NATO and Norway. The topic was "Oil Spill Response – Equipment for Oil Spill Response Operations in the Barents Sea (Cold, Arctic Climate)". The main goal was evaluation of possible foundations of a NRC-CCMS joint developmental project for oil spill response equipment. This goal was linked to environmental protection along shipping lanes to US and Europe used in transport of Russian crude and refined oil.

As a result of this Workshop, it was agreed to establish a working group for further assessment of an upgrade/modification of existing mechanical oil spill response equipment and an assessment of using *in situ* combustion as an operational tool in the Barents Sea. There was a focus too on dispersants.

The second workshop, a follow up on workshop one, took place in Moscow in October 2005 in the framework of NRC. The focus was on "Regulatory and Legal Framework in the Field of Prevention, Localization and Response to Oil Spills on Sea and in Coastal Zones, Review of Methods and Means for Oil Spill Response".

There were 65 experts drawn from Russia, NATO, Norway, Turkey and the USA. Outcomes included:

- Research into environmental clean up of oil spills in the cold Arctic climate during extraction and transport of hydrocarbons
- Computer simulation, forecasting, risk assessment, and investment issues
- New technologies and equipment for liquidation of oil spills
- Use of laser radiation and remote sensing on oil contamination

Further outcomes included an analysis of legal and normative basis of Russia, Norway and Canada in environmental protection during exploration and extraction of hydrocarbons in the Arctic sea shelf. Insurance issues were also addressed.

The four themes of this third workshop centre on *Oil Spill Fate and Transport, Biological Effects, Risk Prevention, and Operational Response*.

When Canada offered to host this NATO Workshop in November 2005 in collaboration with Russia, the NATO Committee responsible was the Committee on the Challenges of Modern Society (CCMS), founded in 1969 under the aegis of the Public Diplomacy Division of NATO. The initial aim of CCMS was to address problems affecting the environment of nations and the quality of life of their peoples. Activities expanded over the years to include partner countries, and the Russian Federation, and progressively to take into account emerging issues of security.

The CCMS provides a unique forum for the sharing of knowledge and experiences on technical, scientific and policy aspects of social and environmental matters, both in the civilian and military sectors. The CCMS program has been adapted to NATO's new mission. Guidelines for future work are centered on five Key Objectives, which are:

1. Reducing the environmental impact of military activities
2. Conducting regional studies including cross-border activities

3.  Preventing conflicts in relation to scarcity of resources

4.  Addressing emerging risks to the environment and society that could cause economic, cultural and political instability and

5.  Addressing non-traditional threats to security

This past summer saw the culmination of an extensive restructuring initiative wherein the CCMS, which tended to operate in bottom-up mode for selecting projects, was merged with the NATO Science Committee (SCOM), which tended to deal with more top-down activities, to form a new Committee entitled now the Committee on Science for Peace and Security (SPS). The need was driven by the rapidly changing global security environment, and the resulting emergence of common priorities in the two previous programs.

The creation of the SPS Committee through a comprehensive restructuring of the Science Committee and the CCMS will result in a simplified, more effective and fully-integrated organization. The focus will be on identifying and rapidly responding to emerging threats and challenges to our societies, and creating linkages with and among NATO's cooperation partners. The SPS Committee will follow a proactive approach to partner country needs by promoting practical cooperation among scientists and experts.

The SPS Committee replaces both the CCMS and the SCOM, bringing together the best aspects of both in a streamlined and more effective new structure. This Workshop is in a sense a swansong for the CCMS. The spirit behind its logo remains the same as it has for its 37 years of existence. Ongoing activities held under the CCMS banner will continue with characteristic vigour under the new SPS umbrella. The SPS meets for its inaugural meeting at NATO, Brussels, on Friday, 20 October 2006. The SPS Committee constitutes the primary NATO committee supporting practical cooperation in civil science and innovation. It maintains a focus on practical visible projects with tangible output.

With those words of introduction, I wish you all a most productive Workshop and an enjoyable stay in the Halifax area.

Dr. Walter Davidson
NATO/CCMS National Representative
Director (National Facilities)
National Research Council of Canada

# INTRODUCTION

Despite improvements in technologies to prevent the spill of crude oil and its refined products in the marine environment, accidental spills will occur in the future. In October 2006, Fisheries and Oceans Canada hosted the 3rd Oil Spill Workshop sponsored by the NATO Committee on Challenges of Modern Society (CCMS). This international workshop provided an open forum for over 110 participants to exchange information, generate new ideas, and form research partnerships that will lead to an improvement of oil spill countermeasure technologies. This book is a direct outcome of the workshop. Participants were invited to submit manuscripts based on the subject of their presentations for publication as a chapter in a book following approval in a peer-review process.

A focal point of this workshop was a special session on oil spill response countermeasures for use under arctic conditions. This is driven by our recognition of increased risk of spills in the future based on emerging evidence of global climate change that may lead to year-round marine transport via the Northwest Passage and the expansion of industrial developments (e.g., offshore oil and gas exploration and production) in the Arctic. Summarized under Part 1 of this book, this session outlined the environmental challenges that must be addressed in oil spill response operations such as harsh cold water conditions, changes in the physico-chemical properties of the oil, the lack of waste disposal sites, ice cover, detection of oil under ice, etc. An overview of traditional techniques for the quantification of oil trapped under ice is provided as well as insights on new state-of-the-art technologies under development. To assist in the selection of the appropriate response tool, an overview of the design and results of meso-scale (large test tank facility) experiments and field trial studies to determine the efficacy and potential human health risks of current oil spill countermeasures such as in-situ burning, the application of chemical herders, and the application of chemical oil dispersants coupled with propeller wash, is provided. The advantages of in-situ countermeasures that do not require physical transport and off-site disposal of large volumes of contaminated materials in the Arctic is obvious. The influence of "weathering" processes and the natural formation of oil mineral aggregates on the environmental persistence of oil spilled under arctic conditions are described in several papers. A description is given for a large scale international effort funded by industry to describe the behavior of oil spilled in the Arctic and the efficacy of various oil spill countermeasures based on the enhancement of natural processes such as physical dispersion.

Part 2 of the book focuses on the application of oil spill countermeasure strategies for use in north-temperate waters. This section describes advances in containment strategies such as the deployment of towed boom arrays, pumping systems, etc. Chemical oil dispersants remain on our priority list for oil spill response at sea. A number of new chemical oil dispersant formulations have been developed in the last decade in response to spills of heavy crude oils that occurred in Europe. The results from a series of sea trials on chemical dispersants conducted in France to elucidate the factors controlling operational success are described in detail. In addition, a description is presented on the strategy developed in Norway for the application of oil dispersants in offshore areas.

Support for chemical oil dispersant use is based on the premise that its application would transport oil on the surface into the water column where it is effectively diluted to concentrations below toxic threshold limits and eventually biodegraded at enhanced rates. A number of chapters are devoted to the study of factors which control the effectiveness of chemical dispersants. Primary factors of interest include the amount of wave energy required for dispersion, the size and stability of oil droplets formed and the influence of suspended particles in the water column such as mineral fines. These factors are being studied under controlled conditions in wave tank facilities that enable quantification of physical and chemical parameters.

Biological effects of hydrocarbons spilled in the marine environment and quantification of subsequent habitat recovery remains a concern to environmental resource managers, oil spill responders and the public. In Part 3, chapters are devoted to the effects of contaminant petroleum hydrocarbons on the organisms living within sediments as well as fish. A major question remains "How clean is clean?" This topic requires further investigation for the development of monitoring tools to provide an operational end-point for remedial activities and is essential for decisions related to com-pensation for environmental damage. Biological (toxicity test assays) and chemical (oil finger-printing and spill source identification) methods are described.

With greater acceptance of chemical oil dispersant use and its proven effectiveness, consideration is being given to the expansion of its operational window to include areas of coastal habitat that were previously not considered in light of concerns over limited dilution rates and the sensitivity of near shore biota. The ecological significance of microbial activity on the degradation of contaminant hydrocarbons is now recognized. Indeed, nutrient addition as a bioremediation (the removal of contaminants by enhancement of biological activity) strategy has been proven to be an effective countermeasure strategy for oil stranded in shoreline environments.

The development of numerical models for the prediction of the fate and transport of oil spilled at sea is summarized in Part 4. The output of such models may be used for the guidance of oil spill response operations and environmental risk assessment studies.

The last section of this book, Part 5, is focused on operational needs for the future including risk assessment, contingency planning, operational response and policy developments. A comprehensive overview of emergency response strategies for several nations (e.g., Belgium, Norway, Canada, United Kingdom, etc.) and regions (e.g., North Sea, Barents Sea, Mediterranean, Black Sea, North Atlantic, St. Lawrence River, etc.) are provided. Methods of risk analysis and decision making tools for the selection of oil spill response tools are described.

It is clearly evident from this book that future improvement of oil spill response strategies will be an international effort as the issue is not defined by national boundaries. All nations have a common goal to protect the habitat of our oceans and the sustainability of its living resources for use by future generations.

Kenneth Lee
Director
Centre for Offshore Oil and Gas Environmental Research
Bedford Institute of Oceanography
Fisheries and Oceans Canada

**PART 1. ARCTIC RESEARCH ISSUES**

# OIL SPILL RESPONSE AND THE CHALLENGES OF ARCTIC MARINE SHIPPING: AN ASSESSMENT BY THE ARCTIC COUNCIL

A. TUCCI[†]

*United States Coast Guard\Office of Response Field Activities Directorate, 2100 2nd St. S.W, Washington, DC 20593-0001, USA*

**Abstract[*].** The Arctic Council has committed to the conduct of an Arctic Marine Shipping Assessment. The state of Arctic climate change, the projections for ice coverage change and the nature of current marine vessel traffic in the Arctic Region indicate that there will be significant potential impacts on the marine environment from the anticipated growth of marine traffic. To address the potential impacts of this growth the Arctic Council has developed a comprehensive initiative to assess with stakeholder participation the growth and challenges to be expected from marine transportation. The Arctic Marine Shipping Assessment includes data acquisition, public and stake holder outreach, and expert analysis in relation to traffic growth, environmental challenges, risk analysis, socio-economic impacts and related matters.

**Keywords:** arctic, shipping, traffic, oil spill response

[†] To whom correspondence should be addressed. E-mail: Andrew.E.Tucci@uscg.mil

[*] Full presentation available in PDF format on CD insert.

W. F. Davidson, K. Lee and A. Cogswell (eds.), *Oil Spill Response: A Global Perspective.*     3
© Springer Science + Business Media B.V. 2008

# OIL SPILLS IN THE ARCTIC: A REVIEW OF THREE DECADES OF RESEARCH AT ENVIRONMENT CANADA

B. HOLLEBONE[†] & M.F. FINGAS
*Emergencies Science and Technology Division,
Science and Technology Branch, Environment Canada,
Environmental Technology Centre, 335 River Rd.,
Ottawa, ON, Canada*

**Abstract\***. Since 1970, Environment Canada has had the responsibility to coordinate response for environmental emergencies in Canada, to develop new understandings of how emergencies happen, their effects on Canada's environment, and to develop and test new techniques to protect the environment from their adverse repercussions. The Arctic and Marine Oil spill Program (AMOP) was initiated by Environment Canada in conjunction with many partners to improve capabilities to detect oil in the Arctic, to understand the fate and behaviour of oil in ice and to counteract and limit the impacts of oil spills in the Arctic and marine environments. For the past thirty years, AMOP has sponsored and participated in hundreds of individual research projects in each of these three fields of research. Finding oil in Arctic waters is made difficult both by the presence of ice and by the long darkness of winters. Environment Canada has developed a second-generation airborne laser fluorosensor (SLEAF), autonomous oil sensor buoys, and new techniques to assess oiled Arctic shorelines (SCAT). The second major focus of AMOP has been to develop understanding of the fate and behaviour of oil in the Arctic. These efforts have included large-scale field projects such as the Baffin Island Oil Spill (BIOS) and the Newfoundland Offshore Burn Experiment (NOBE), laboratory studies of oil in ice (for example, the effects of oil on ice growth, the degradation of oil in ice, and the sinking of oil in cold water), the development of new analytical environmental forensic techniques to identify, quantify, and understand the behaviour of oil, and the creation of one of

---

[†] To whom correspondence should be addressed. E-mail: bruce.hollebone@ec.gc.ca

\* Full presentation available in PDF format on CD insert.

W. F. Davidson, K. Lee and A. Cogswell (eds.), *Oil Spill Response: A Global Perspective.*     5
© Springer Science + Business Media B.V. 2008.

the largest catalogues of oil properties at cold temperatures. Environment Canada has also been working to improve spill countermeasures in the Arctic. This has included the development and testing of new equipment including ice skimmers, conventional booms, water jet barriers, fire-resistant booms and slick igniters. Other countermeasures are also actively researched including the burning of oil and bitumens in ice, and the use of dispersants and other spill-treating agents. Finally, to communicate with and to bring together oil researchers from around the world, AMOP conducts an annual seminar (with peer-reviewed proceedings) now in it's 29th year, which has developed into one of the major technical conferences focusing on oil spill detection, behaviour and countermeasures.

**Keywords:** arctic, oil spill response, environment, oil in ice, AMOP

# OIL UNDER ICE DETECTION:
# WHAT IS THE STATE-OF-THE-ART?

RON GOODMAN[†]

*Innovative Ventures (ivl) Ltd., Cochrane, Alberta, T4C 1A8, Canada*

**Abstract.** Since the exploration for oil and gas in the Canadian and US arctic commenced in the early 1970s, a need has been identified to develop technology to detect oil under ice. Both electromagnetic and acoustic sensors have been tried, but a practical field instrument has not been identified. Most proposed systems require that the equipment be operated from the ice surface in order to get adequate coupling and, for some systems, the snow must be removed from the ice. For many ice situations, surface access is difficult and poses a severe safety issue. Two recent spills in Alberta used "high technology" ice augers to detect the presence of oil under the ice. Some potential new techniques are discussed and the basic principles of their operation described.

**Keywords:** arctic, oil spill response, oil in ice, detection

## 1. Introduction

The detection of oil under continuous ice cover has presented one of the most difficult challenges to the oil-spill technological community for the past two decades and there is still no operationally proven system available. Dickins (2000) under the sponsorship of the US Minerals Management Service conducted an excellent review of the status of oil-under-ice detection and this paper complements this review with a more detailed analysis of some systems. Dickins identified many false start concepts, which will not be discussed in this paper. In order to determine the design of a suitable oil-under-ice detector, the various situations under which oil may be found under a continuous ice sheet need to be considered.

---

[†] To whom correspondence should be addressed. E-mail: goodmanr@cia.com

W. F. Davidson, K. Lee and A. Cogswell (eds.), *Oil Spill Response: A Global Perspective.*    7
© Springer Science + Business Media B.V. 2008

The oil must come from a sub-surface release since any surface release would either be on the ice surface or in a lead or other opening in the ice. Potential sources of sub-surface oil are a leak in a pipeline, the leakage from a submerged tank or vessel or a natural seep. Oil when trapped under ice does not spread rapidly or cover a large area due to natural roughness of the ice-water interface (Rosenegger, 1975). The situation is analogous to oil spilled on land, rather than the more dynamic situation of oil on water. Unlike the oil-on-water situation, the probable location of the source of the oil can be well defined spatially, so the search for the oil is over a relatively small confined area. Depending on the time of year, the ice may just be forming, be in a rapid growth phase, be essentially static or in a break-up situation, so that the oil may be on the surface surrounded by ice floes, at the ice-water interface or in corporated in the growingice sheet. In the firstand last case, traditional remote sensing techniques can be used to detect the oil. When the oil is at the ice-water interface or incorporated in the ice sheet, new oil-under-ice detection systems are required (Figure 1). The basic mode of detection may be different for the two situations.

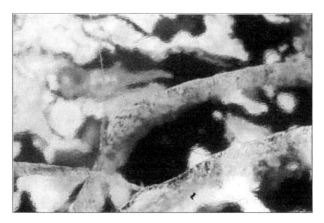

*Figure 1.* Oil encapsulated in fresh-water ice at Norman Wells, NWT.

What would the specifications, in order of priority, of an ideal oil-under-ice detector be?

- Detect small amounts of oil under 2 m of ice and variable snow depths with no false positives.

- Detect oil lenses in the ice at a size scale of a few centimeters in thickness of an area of less than 1 m$^2$.

- Determine oil thickness; but this may be an illusion since the thickness will spatially vary due to ice roughness.

- Operate from an airborne platform at altitudes of greater than 300 m under a wide range of weather conditions. Since in most cases the location of oil under ice will be somewhat localized this may not be an important criteria. The remote sensing requirement may be more critical in terms of safety issues associated with working from the ice surface.

- Produce a georeferenced map in near real time.

There is no system that can meet these criteria and only three systems show any potential for meeting the first two, which are the most critical.

In order to evaluate the various oil-under-ice detection systems, the acoustic and electromagnetic properties of the various components of the system are required and are summarized in Table 1, from Goodman *et al.* (1985) and Langleben and Pounder (1970).

TABLE 1. Acoustic and electromagnetic properties of air, ice, oil and water.

| Material | Acoustic impedance | Dielectric constant |
|----------|-------------------|---------------------|
| Air      | 43                | 1                   |
| Ice      | 300,000           | 3.5                 |
| Oil      | 119,000           | 2–3                 |
| Water    | 148,300           | 80                  |

Using the equation (Mott, 1971):

$$R = \left| \frac{E_r}{E_i} \right|^2 = \left( \frac{n_2 - n_1}{n_2 + n_1} \right)^2 \quad (1)$$

where:

$R$ = Proportion of energy reflected
$E_r$ = Reflected energy
$E_i$ = Incident energy

$n_1$ = Dielectric constant or acoustic impedance of the incident medium

$n_2$ = Dielectric constant or acoustic impedance of the second medium.

Based on this simple equation, the reflection of acoustic energy is total with an acoustic signal originating in the air into any of the other three media. The reflection of electromagnetic radiation is low into ice or oil from the air, but nearly total in the case of water (Figure 2).

*Figure 2.* Measuring oil under ice roughness in the Canadian Arctic.

## 2. Existing Technology

The signal associated with the detection of oil under ice may be due to dielectric or acoustic impedance difference between the oil and the ice, or by a change in the surface roughness of the oil-ice interface. The rougher the interface, the more the probing signal is scattered and hence the weaker the signal returned to the receiver. This is the basis, for example, of the detection of oil-on-water using radar. The interface roughness has been directly measured using a mould system deployed by divers (Goodman *et al.*, 1987) and found to be rough at spatial scales of meters and roughness values of several centimetres. The oil released under ice fills the roughness features and generates a smooth interface with the water, which can be detected using either acoustic or electro-magnetic sensors.

## 2.1. MECHANICAL SYSTEMS

The only proven and widely used technology is to drill a hole in the ice using an ice auger, a chain saw or similar mechanical system. While this is time consuming and is a single point measurement, it works. In order to increase the productivity of such units, they can be mounted on a small snow vehicle to increase their coverage, but consideration must be given to the additional safety concerns of using such equipment on ice sheets of unknown thickness. Using hand-held systems, ice thick-nesses greater than about a meter and a half are difficult (Figure 3).

*Figure 3.* Ice auger at Lake Wabamun, Alberta, Canada. Photo credit Pat Lambert.

Some experiments (Dickins *et al.*, 2005) have been conducted on the detection of the vapour from the oil that would permeate through the ice and be trapped on the surface. While this system worked well in the laboratory environment, it would be very difficult to implement in a typical cold weather environment. This system is very time consu-ming to install and the time for each measurement took several minutes. There is some evidence from field experiments that very little evapor-ation occurs under an ice sheet, so the presence of vapours in the labo-ratory experiment may well be an artefact of the experimental situation (Figure 4).

*Figure 4*. Plastic dome used for the collection of oil vapours. From Dickins *et al.*, 2005.

## 2.2. ELECTROMAGNETIC SENSORS

The electromagnetic band extends from long-wavelength radio waves
to X-rays, and includes the visible band and radar. Various parts of the
electromagnetic spectrum had been tried for the detection of oil under
ice, including low-frequency systems at about 100 kHz, various forms
of radar from 100 to 1,000 MHz, and the visible band either directly or
by detecting the fluorescence of the oil. While there is some variation
of the dielectric constant with frequency, the values of Table 1 are
typical. It is easy for electromagnetic radiation to be transmitted from
the air to either the ice or the oil.The reflection at the oil-ice interface
will be weak, but easily detected provided the sensor has an adequate
dynamic range. There will be a strong reflection at the interface with
the water. As with any sensing package, the spatial resolution depends
on the wavelength (and pulse length for pulsed systems).

There are a number of low-frequency electromagnetic systems, which
use induction to detect surface and sub-surface anomalies (Figure 5).
These systems typically operate at frequencies below 100 kHz (wavelengths of greater than 3,000 m). At these large wavelengths the spatial
resolution is poor and while such systems have been proven useful for
sea-ice thickness measurements, it is unlikely that this group of sensors
would detect oil either in ice or at the ice-water interface (Kovacs
*et al.*, 1995).

*Figure 5*. Electromagnetic induction sensor. Photo by Fugro Airborne Surveys. Source unknown.

Ground penetrating radar (GPR) systems are routinely used to determine sub-surface structures and operate at frequencies between 300 and 1,000 MHz (Figure 6). (1 m to 30 cm wavelengths). In order to achieve good spatial resolution, most GPR systems use a high-bandwidth antenna (low Q) and produce a short chirp signal (Moorcroft and Tunaley, 1985). Most of the currently available GPR systems are surface based and require good coupling between the unit antennae and the ice. Since such systems are routinely commercially available, they are very attractive to be used as an oil-under-ice detector. There have been a number of experiments, both in test basins and in the field to test the ability of these systems to uniquely detect oil-under-ice. The main problem is both signal strength and dynamic range, since, depending on the value of the dielectric constant of the oil, the reflected signal difference between oil and ice is 0.5–7% as opposed to the nearly 100% at the ice-water interface. Thus, the receiver must be sensitive to small variations in signal strength to see the oil-ice interface, while not being overloaded by the return from the water. Older sys-tems lacked this

*Figure 6*. Ground Penetrating Radar (GPR) system in the Arctic. Source unknown.

dynamic range and the ice-oil signal was masked by the water return. The electronics used by more recent designs have a better dynamic range, and Dickins *et al.* (2005) have recently used such a system to evaluate an oil-under-ice detection in a test basin using urea ice, and subsequently (Brandvik *et al.*, 2006) in an experiment in the Norwegian Arctic. The test basin experiments used extensive signal analysis in order to identify the presence of oil under ice, which obscures what properties of the interface are actually being detected. The use of urea ice, whose electrical properties are different from natural ice, further complicates the interpretation. The field data from 2006 is still being analyzed.

For more than a decade, radio-echoing sounding systems operating in the same frequency band as the GPS have been used to measure glacier ice thickness from an airborne platform (Figure 7). These systems have a much narrower bandwidth and beam width than a typical GPS system, but offer the potential to remotely detect oil under ice. Since these systems have a much lower spatial resolution than the GPS, a larger area of oil-under-ice would be required for a reasonable test of the units' capability. No such field studies have been undertaken.

*Figure 7.* British Antarctic survey Twin Otter with radio echo sounding system Strasbourg, France.

The most readily available electromagnetic sensor is the human eyeball, and this can be used either directly or to detect ultraviolet induced fluorescence of the oil (Moir and Yetman, 1993). The photo detector may be as simple as the human eye protected with goggles from the incident UV excitation, or a more complex electronic detection system. The only benefit of the latter is that the output may be quantified and recorded. The laser fluorosensor operates on a similar basis but can function from an airborne platform. Of all the electromagnetic techniques, UV fluorescence is the only one that can uniquely identify the presence of oil (Figure 8). The simplest method of oil detection under ice is to use a subsurface illumination and observe the oil slick directly (Figure 9). This of course requires the use of some mechanical means to penetrate the ice and insert the waterproof lamp. An example of this technique is shown in Dickins *et al.* (2005). This will work well for ice thicknesses of up to a meter, but requires that the ice surface be snow free.

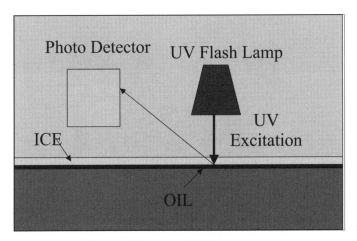

*Figure 8.* Schematic of UV fluorescence.

## 2.3. ACOUSTIC SYSTEMS

As can be seen from Table 1, acoustic energy is completely reflected from the air to ice or air to water interface, so any acoustic sensor must be tightly coupled to the ice surface and will probably require either antifreeze or some other liquid to ensure good coupling between the transducer and the ice. The return signal from the ice-liquid interface will be stronger than for the electromagnetic case, the difference in returns from the water and oil will be similar in intensity.

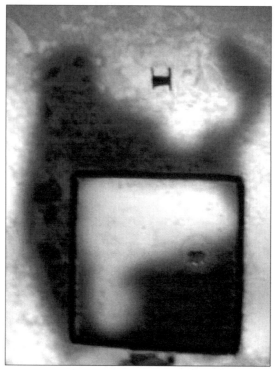

*Figure 9.* Oil under ice detection using sub-surface illumination. From Dickins *et al.* (2005).

However, acoustic signals can be propagated in two different modes: shear (S) (velocity 1,900 m/s) and compressive or pressure (P) (velocity 3,980 m/s) waves (Jones *et al.*, 1986) (Figure 10). While it is possible to develop specialized detectors for these different types of waves, they propagate in ice at very different velocities, and thus the detection process can be greatly simplified to a time domain analysis. Acoustically, at high frequencies, the oil appears as a semi-solid, and can propagate both shear and pressure waves. For an ice-water interface, only compressive waves will be returned, so the presence of two return peaks uniquely identifies oil at the ice-water interface. The choice of frequency of operation is a compromise between spatial resolution (needing higher frequencies) and depth penetration (needing lower frequencies). For the prototype, a frequency of 100 kHz was chosen.

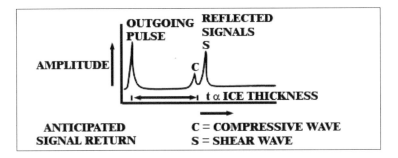

*Figure 10.* Acoustic return signals showing shear and compressive waves.

This system was initially tested in the Esso test basin in Calgary, followed by tests using the natural oil seeps in the Mackenzie River at Norman Wells and in the Canadian Beaufort. In all cases the oil was readily detected under the ice if there was a signal return. In some cases no return was received due to problems with coupling of the acoustic system to the ice or where the ice was shattered near the shore. Figure 11 shows the system being tested in the Canadian Beaufort Sea. The small cylinders are the transmitters and the receiver is the multi-detector array. The oil was placed in a small containment boom under the ice by divers. This system was developed in the 1980s using the computer

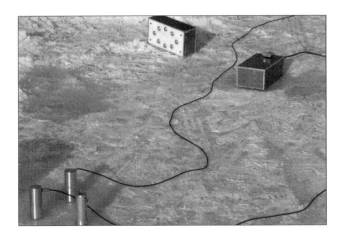

*Figure 11.* Acoustic system in the Canadian Arctic.

and electronic technology available at that time. It is clear that the system works, but there have not been sufficient funds available to upgrade the system to use currently available computing technology.

## 3. What's Next?

Most of the oil-under-ice detection work has focused on the use of sensors from the surface, thus avoiding the need to put a hole in the ice. The acoustic properties of the materials involved clearly suggest that an in water sensing system would work very well and provide unambiguous detection of the presence of oil under ice, and an even simpler system would use a submersible mounted light and video camera (McKindra *et al.*, 1981).

## 4. Conclusions

The only proven technology is the use of mechanical equipment to make a hole in the ice and visually detect the presence of oil. This can be augmented using a sub-surface light to see the oil in many situations. While this may seem difficult, if any response is to be undertaken, the ice surface must be breeched.

Both radar and acoustic systems show promise, but still need development. The acoustic systems are preferred since they can discriminate between oil and the presence of ice layers, which confuse the interpretation of the return.

## References

Brandvik, P.J., Faksness, L.-G., Dickins, D., and Bradford, J., 2006, Weathering of Oil Spills Under Arctic Conditions: Field Experiments with Different Ice Conditions Followed by In-situ Burning, in: *Proceedings of the 3rd Annual NATO/CCMS Oil Spill Response Workshop, Halifax, NS, Canada.*
Dickins, D.F. *Detection and Tracking of Oil Under Ice*, Final Report, DF Dickins Associates Ltd, Escondido, California, October 6, 2000, Minerals Management Service, Report 348.
Dickins, D.F., Bradford, J., Liberty, L., Hirst, B., Owens, E., Jones, V., Gibson, G., Zabilansky, L., and Lane, J., 2005, Testing Ground Penetrating Radar and Ethane

Gas Sensing to Detect Oil in and Under Ice, in: *Proceedings of 28th Arctic and Marine Technical Seminar, Environment Canada, Ottawa, ON, Canada,* pp. 799–824.

Goodman, R., Dean, A., and Fingas, M., 1985, Detection of Oil Under Ice Using Electromagnetic Radiation, in: *Proceedings 8th International Conference on Port and Ocean Engineering Under Arctic Conditions, Narssarssuaq, Greenland, September 7–14, 1985, II* pp. 895–902.

Goodman, R., Holoboff, A., Daley, T., Waddell, P., Murdock, L. and Fingas, M., 1987, A Technique for the Measurement of Under-ice Roughness to Determine Oil Storage Volumes, in: *Proceedings, 10th International Oil Spill Conference, Baltimore, MD,* pp. 395-398.

Jones, H., Kwan, H., and Yeatman, E., 1986, The Detection of Oil Under Ice by Ultrasound Using Multiple Element Phased Arrays, in: *Proceedings 9th annual AMOP Technical Seminar, Environment Canada, Ottawa, ON, Canada,* pp. 475–484.

Kovacs, A., Holladay, J.S., Bergeron, Jr. C.J., 1995, The footprint/altitude ratio for helicopter electromagnetic sounding of sea-ice thickness: comparison of theoretical and field estimates. *Geophysics,* **60**(2): 374–380.

Langleben, M.P. and Pounder, E.R., 1970, *Acoustic Attenuation in Sea Ice and Reflection of Sound at the Water-Ice Interface.* Report 816, MacDonald Physics Lab, McGill University, 30 pp.

Mott, G., 1971, Reflection and refraction coefficients at fluid-solid interface. *Acoust. Soc. Am.* **50**, 819–829.

McKindra, C.D., Dunton, K.H., and Karns, R., 1981, Electronic Diver (Under Ice) Detection System, in: *Proceedings 4th annual AMOP Technical Seminar, Environment Canada, Ottawa, ON, Canada,* pp. 569–586.

Moir, M.E. and Yetman, D.C., 1993, The Detection of Oil Under Ice by Pulsed Ultraviolet Fluorescence, in: *Proceedings 1993 International Oil Spill Conference, Tampa, FL,* pp. 521–523.

Moorcroft, R. and Tunaley, J., 1985, Electromagnetic Resonance in Layers of Sea Ice and Oil Over Sea Water, in: *Proceedings 8th annual AMOP Technical Seminar, Environment Canada, Ottawa, ON, Canada,* pp. 269–286.

Rosenegger, L.W., 1975, *The Movement of Oil Under Sea Ice.* Prepared by Imperial Oil Limited for the Beaufort Sea Project Report #28, Environment Canada, Victoria, BC.

# PRACTICAL AND EFFECTIVE TECHNOLOGIES FOR OIL SPILL RECOVERY OPERATIONS IN ARCTIC CONDITIONS

J.P. MACKEY[†]

*Lamor Corporation LLC, 28045 Ranney Parkway, Cleveland, OH 44145, USA*

**Abstract**[*]. There is a growing awareness of the need for reliable and effective tools to fight oil spills in Arctic conditions. Extreme cold water and air temperatures affect the functioning and reliability of machinery and change the physical characteristics of spilled oil, often rendering traditional recovery methods ineffective. The presence of ice affects the operation of floating equipment as well as the behaviour and concentration of oil on the water surface. Responders in Finland and other Baltic countries are faced with these operational challenges of extreme cold and ice, while at the same time, the risk of oil spills from facilities and vessels is increasing at an unprecedented rate due to expanding oil export from Russia This has stimulated intense development of new technologies and techniques which can benefit all responders in cold regions of the world. This presentation will discuss stiff brush oil recovery technology, which has proven to be a reliable and highly effective tool in these harsh environments. Several new and promising machines will be reviewed. Emphasis will be on practical design principles and techniques that are being used to optimize performance of these systems.

**Keywords:** oil spill recovery, arctic, ice, technology

---

[†] To whom correspondence should be addressed. E-mail: Jim.Mackey@Lamor.com
[*] Full presentation available in PDF format on CD insert.

W. F. Davidson, K. Lee and A. Cogswell (eds.), *Oil Spill Response: A Global Perspective.*     21
© Springer Science + Business Media B.V. 2008

# *IN-SITU* BURNING FOR OIL SPILLS IN ARCTIC WATERS: STATE-OF-THE-ART AND FUTURE RESEARCH NEEDS

STEPHEN POTTER[†] & IAN BUIST

*SL Ross Environmental Research, 200-7 17 Belfast Rd., Ottawa, Ontario, Canada K1G 0Z4*

**Abstract.** *In-situ* burning is one of the few practical options for removing oil spilled in ice-covered waters. In many instances *in-situ* burning, combined with surveillance and monitoring, may be the only response possible. As with all countermeasures in any environment, the suitability of burning a particular spill depends on the characteristics of the spilled oil and how the oil behaves in the particular ice conditions. There is an extensive body of knowledge concerning *in-situ* burning of oil in ice situations, beginning with laboratory, tank and field studies in the mid-1970s in support of drilling in the Canadian Beaufort Sea. *In-situ* burning research has been conducted primarily in Canada, Norway and the United States. This paper serves as a review of the subject, incorporating recent research results, summarizing the following topics:

- The basic requirements and processes involved with *in-situ* burning
- Trade-offs associated with burning in ice-covered waters
- How oil spill behavior in various ice conditions controls *in-situ* burning
- The application of burning in various common ice situations and
- Key equipment requirements.

**Keywords:** *in-situ* burning, oil spill response, arctic

---

[†] To whom correspondence should be addressed. E-mail: Steve@slross.com

W. F. Davidson, K. Lee and A. Cogswell (eds.), *Oil Spill Response: A Global Perspective.*   23
© Springer Science + Business Media B.V. 2008

## 1.  Introduction

The use of *in-situ* burning as a spill response technique is not new, having been researched and employed in one form or another at a variety of oil spills since the late 1960s. *In-situ* burning is especially suited for use in ice conditions, often offering the only practical option for removal of surface oil in such situations. Much of the early research and development on *in-situ* burning focused on its application to spills on and under solid sea ice. Most recently, the research has addressed burning spills in loose pack ice. In general, the technique has proved very effective for thick oil spills in high ice concentrations and has been used successfully to remove oil resulting from pipeline, storage tank and ship accidents in ice-covered waters in Alaska, Canada and Scandinavia (Buist *et al.*, 1994; Guénette, 1997).

Although there have been numerous incidents of ship and oil well spills that inadvertently caught fire, the intentional ignition of oil slicks on open water has only been seriously considered since the development of fire-resistant oil containment booms beginning in the early 1980s.

The development of these booms offered the possibility of conducting controlled burns in open water conditions. *In-situ* burning operations using these booms have been conducted at three open water spills in North America in the 1990s: a major offshore tanker spill, a burning blowout in an inshore environment, and a pipeline spill into a river. The new generations of fire containment booms presently available commercially represent a mature technology: the best have been subjected to standardized testing that verifies their suitability and durability.

*In-situ* burning of thick, fresh slicks can be initiated very quickly by igniting the oil with devices as simple as an oil-soaked sorbent pad. *In-situ* burning can remove oil from the water surface very efficiently and at very high rates. Removal efficiencies for thick slicks can easily exceed 90%. Removal rates of 2,000 $m^3$/h can be achieved with a fire area of only about 10,000 $m^2$ or a circle of about 100 m in diameter. The use of towed fire containment boom to capture, thicken and isolate a portion of a spill in low ice concentrations, followed by ignition, is far less complex than the operations involved in mechanical recovery, transfer, storage, treatment and disposal. If the small quantities of residue from an efficient burn require collection (research indicates that burn residue is of low acute toxicity to marine organisms, but may smother benthic resources if it sinks), the viscous, taffy-like material can be collected and stored for further treatment and disposal. There is a limited window of opportunity for using *in-situ* burning with the presently available technology. This window is defined by the time it takes the oil slick to emulsify; once water contents of stable emulsions exceed about 25%, most slicks are unignitable. Research has shown how it may be possible to overcome this limitation by spraying the slick with demulsifying chemicals.

Despite the strong incentives for considering *in-situ* burning as a primary countermeasure method, there remains some resistance to the approach. There are two major concerns: first, the fear of causing secondary fires that threaten human life, property and natural resources; and, second, the potential environmental and human-health effects of the by-products of burning, primarily the smoke.

The purpose of this paper is to review the science, technology and ecological consequences of *in-situ* burning as a countermeasure for oil spills in ice conditions. The main focus is on marine oil spills; however, spills in snow are also covered (since many spills on ice will inevitably involve snow). Much of the content of this chapter is adapted from: an in-depth review of *in-situ* burning produced for the Marine Spill Response Corporation (MSRC) (Buist *et al.*, 1994) summarized and updated for IUPAC (Buist *et al.*, 1999) and the USCG In-situ Burn Operations Manual (Buist et al., 2002). Interested readers are encouraged to refer to the original reports for fully referenced details of the sum-mary presented here. The MSRC report is available from the American Petroleum Institute in Washington, DC and the USCG Manual is available from the USCG R&D Center in Groton, CT. Both documents are contained on a CD produced by NIST for MMS that contains a large number of the key references on *in-situ* burning (Walton and Mullin, 2003).

## 2.    The Fundamentals of *In-Situ* Burning

### 2.1.    REQUIREMENTS FOR IGNITION

In order to burn spilled oil, three elements must be present: fuel, oxygen and a source of ignition. The oil must be heated to a temperature at which sufficient hydrocarbons are vaporized to support combustion in the air above the slick. It is the hydrocarbon vapors above the slick that burn, not the liquid itself. The temperature at which the slick produces vapors at a sufficient rate to ignite is called the Flash Point. The Fire Point is the temperature a few degrees above the Flash Point at which the oil is warm enough to supply vapors at a rate sufficient to support continuous burning. The essential elements of a burning pool of liquid are:

- That the liquid is heated to its Fire Point.
- Once at its Fire Point, the liquid evaporates quickly.
- Vapors from the liquid burn in the air above the pool.
- Air for combustion is drawn in by the plume of rising combustion gases.

## 2.2. IMPORTANCE OF SLICK THICKNESS

The key oil slick parameter that determines whether or not the oil will burn is slick thickness. If the oil is thick enough, it acts as insulation and keeps the burning slick surface at a high temperature by reducing heat loss to the underlying water. This layer of hot oil is called the "hot zone". As the slick thins, increasingly more heat is passed through it; eventually enough heat is transferred through the slick to drop the temperature of the surface oil below its Fire Point, at which time the burning stops.

## 2.3. EFFECT OF EVAPORATION ON SLICK IGNITION

Extensive experimentation on crude and fuel oils with a variety of igniters in a range of environmental conditions has confirmed the following "rules-of-thumb" for relatively calm, quiescent conditions:

- The minimum ignitable thickness for fresh crude oil on water is about 1 mm.
- The minimum ignitable thickness for aged, unemulsified crude oil and diesel fuels is about 2–5 mm.
- The minimum ignitable thickness for residual fuel oils, such as IFO 380 (aka Bunker "C" or No. 6 fuel oil) is about 10 mm.
- Once a 1 m$^2$ of burning slick has been established, ignition can be considered accomplished.

## 2.4.  OTHER FACTORS AFFECTING IGNITION

Aside from oil type, other factors that can affect the ignitability of oil slicks on water include: wind speed, emulsification of the oil and igniter strength. The maximum wind speed for successful ignition of large burns is 10–12 m/s. For weathered crude that has formed a stable water-in-oil emulsion, the upper limit for successful ignition is about 25% water. Some crudes form meso-stable emulsions that can be easily ignited at much higher water contents. Paraffinic crudes appear to fall into this category.

Secondary factors affecting ignitability include ambient temperature and waves. If the ambient temperature is above the oil's flash point, the slick will ignite rapidly and easily and the flames will spread quickly over the slick surface; flames spread more slowly over oil slicks at sub-flash temperatures.

## 2.5.  OIL BURNING RATES

The rate at which *in-situ* burning consumes oil is generally reported in units of thickness per unit time (mm/min is the most commonly used unit). The removal rate for *in-situ* oil fires is a function of fire size (or diameter), slick thickness, oil type and ambient environmental conditions. For most large (>3 m diameter) fires of unemulsified crude oil on water, the "rule-of-thumb" is that the burning rate is 3.5 mm/min. Automotive diesel and jet fuel fires on water burn at a slightly higher rate of about 4 mm/min.

## 2.6.  EFFECTS OF EMULSIFICATION

Although the formation of water-in-oil emulsions is not as dominant a weathering process with spills in ice as it is for spills in open water, emulsions could be formed in some situations (i.e., a subsea blowout). Emulsification of an oil spill negatively affects *in-situ* ignition and burning. Emulsion water contents are typically in the 60–80% range with some up to 90%. The oil in the emulsion cannot reach a temperature higher than 100°C until the water is either boiled off or removed. The heat from the igniter or from adjacent burning oil is used mostly to boil the water rather than heat the oil to its fire point.

A two-step process is likely involved in emulsion burning: "breaking" of the emulsion, or possibly boiling off the water, to form a layer of unemulsified oil floating on top of the emulsion slick; and, subsequent combustion of this oil layer. High temperatures are known to break emulsions. Surface-active chemicals called "emulsion breakers", common in the oil industry, may also be used.

For stable emulsions the burn rate declines significantly with increasing water content, with 25% water content being the upper limit for effective burning for most emulsions. (There are exceptions: some crudes form meta-stable emulsions that can be burn at much higher water contents. Paraffinic crudes appear to fall into this category).

## 3.   Environmental and Human Health Risks

This section describes the main risks associated with *in-situ* burning of spills and the safety measures used to overcome these risks. Much of the material in this section was developed for open water burn operations with towed fire boom, but is also applicable to burns in ice-covered waters. Humans and the environment may be put at risk by:

- The flames and heat from the burn
- The emissions generated by the fire and
- The residual material left on the surface after the fire extinguishes.

### 3.1. FIRE AND HEAT

Flames from *in-situ* burning pose a risk of severe injury or fatality to both responders and wildlife. The threat is obvious and needs no elaboration. This section, then, focuses on the problem of the heat radiated by the burn. Risks exist both in normal operations and abnormal conditions such as tow vessel breakdown and boom failure. The risk to spill responders at the spill site is the main concern because the risks to the general public will be eliminated through the use of an exclusion zone surrounding the spill site.

### 3.2. EFFECTS OF HEAT ON SPILL RESPONDERS

*In-situ* burning of oil produces a large amount of heat that is transferred into the environment through convection and radiation. About 90% of the heat generated by *in-situ* combustion is convected into the atmosphere. The remainder is radiated from the fire in all directions, but there is most concern with heat radiated towards responders, causing heat exhaustion and burns to unprotected skin. Of lesser concern is heat transferred downward which might affect water column resources. The potential for causing injury to exposed workers is a function of both the level of incident radiation and the duration of exposure. Wood will char if positioned about half a fire diameter from the edge of an oil burn. The "safe approach distance" to an *in-situ* oil fire for a person is from two to four times the diameter of the fire depending on the duration of exposure. Conservatively, it is assumed that the safe approach distance to the edge of an *in-situ* oil fire is approximately four fire diameters.

It is important to recognize that the oil contained in a towed boom is relatively thick in the early stages of a burn and that this thick-ness is maintained through towing. If the towing were to stop or slow, or the boom were to break, this thick layer would spread quickly to cover an area several times that of the boomed oil. This will increase the fire diameter, the heat flux from the fire, and the need for workers to move further from the fire to avoid discomfort.

## 3.3.  ENVIRONMENTAL EFFECTS OF HEAT

Heat from the flames is radiated downward as well as outward and much of the heat that is radiated downward is absorbed by the oil slick. Most of this energy is used to vaporize the hydrocarbons for further burning, but a portion of the heat is passed to the underlying water. In a towed-boom burn or in a stationary boom situation in current, the water under the slick does not remain in contact with the slick long enough to be heated appreciably. However, under static conditions (the slick does not move relative to the underlying water – for example in a melt pool) the upper layer of the underlying water may be heated in the latter stages of the burn. In a prolonged static burn, the top few millimeters of the water column may be heated to near boiling temperatures, but the water several centimeters below the slick has been proven to be unaffected by the fire. As a result, the environmental impact of the heat from an *in-situ* burn is likely to be negligible.

3.4. AIR EMISSIONS

The smoke plume emitted by a burning oil slick on water is the main concern. The concentrations of smoke particles at ground or sea level are of concern to the public and they can persist for a few miles downwind of a burn. The smoke plume is composed primarily of small carbon particles and combustion gases. Smoke particles cause the greatest risk in a plume. Carbon smoke particles are responsible for providing the characteristic black colour of the plume rising from a burn. The smoke is unsightly but more important; the smoke particles can cause severe health problems if inhaled in high concentrations. Smoke particulates and gases; however, are quickly diluted to below levels of concern. The amounts of PAHs in the smoke plume are also below levels of concern. Approximately 5–15%, by weight, of the oil burned is emitted as smoke particles.

Descriptions of the constituents of the smoke plume and their dissipation under various conditions can be found in Buist *et al.* (1994). Suffice to say here that although the smoke that is generated can be a dramatic and visible effect from any large *in-situ* burn, the smoke plume is sent high in the air by the hot combustion gases, such that sea-level concentrations of soot are below levels of concern from 3 to 6 km (2–4 nautical miles) downwind.

3.5. BURN RESIDUE

As a general rule of thumb, the residue from an efficient burn of crude oil on water is environmentally inert. More specifically, the potential environmental impacts of burn residues are related to their physical properties, chemical constituents and tendency to float or submerge. Correlation between the densities of laboratory-generated burn residues and oil properties predict that burn residues will submerge in sea water when the burned oils have:

- Initial density greater than 0.865 $g/cm^3$ (API gravity less than about 32°) or
- Weight percent distillation residue (at >540°C) greater than 18.6%.

Burn residues usually submerge only after cooling. Based on modeling the heat transfer, it is likely that the temperature of a 1-cm thick burn residue will reach that of ambient water within approximately 20–30 min. Even for thicker slicks, it is likely that this cooling would occur within approximately 2 h.

Physical properties of burn residues depend on burn efficiency and oil type. Efficient burns of heavier crudes generate brittle, solid residues (like peanut brittle). Residues from efficient burns of other crudes are described as semi-solid (like cold roofing tar). Inefficient burns generate mixtures of unburned oil, burned residues and soot that are sticky, taffy-like or liquid. Burns of light distilled fuels result in a residue that is similar to the original fuel but contains precipitated soot.

## 4.   Burning Spills in Ice and Snow

*In-situ* burning has been considered as a primary Arctic spill counter-measure since before the start of offshore drilling in the Canadian Beaufort Sea in the mid-1970s. Field trials at that time demonstrated that on-ice burning offered the potential to remove almost all of the oil present on the surface of landfast ice with only minimal residue volumes left for manual recovery. This area of research culminated in 1980 with a full-scale field research program on the fate and cleanup of sub-sea oil well blowouts under landfast sea ice.

### 4.1.  SPILLS IN ICE

Research in oil spill cleanup in pack, or broken, ice also began in the 1970s. Interest in the subject increased in the early 1980s because of proposals for offshore production in Alaska and Canada, and has become an international subject of R&D with the opening of Russian ice-covered waters for exploitation and the future potential for drilling in Norwegian ice-covered seas. Interest in the subject has been re-kindled in Alaska with several recent offshore development proposals near Prudhoe Bay. Also, operators of established production facilities in Cook Inlet have an ongoing need to improve their level of under-standing of alternative response strategies for spills in broken ice.

The consensus of the research to date on spill response in broken ice conditions is that *in-situ* burning is a suitable response technique, and in many instances may be the only cleanup technique applicable (Shell *et al.*, 1983; SL Ross, 1983; SL Ross and DF Dickins, 1987).

A considerable amount of research was done on the potential for *in-situ* burning in broken ice, including several smaller-scale field and tank tests (Shell *et al.*, 1983; Brown and Goodman, 1986; Buist and Dickins, 1987; Smith and Diaz, 1987; Bech *et al.*, 1993; Guénette and Wighus, 1996; SL Ross and DF Dickins, 2003). Most of these tests involved large volumes of oil placed in a static test field of broken ice resulting in substantial slick thicknesses for ignition. The few tests in unrestricted ice fields or in dynamic ice have indicated that the efficacy of *in-situ* burning is very sensitive to ice concentration and dynamics (and thus the tendency for the ice floes to naturally contain the oil), the thickness (or coverage) of oil in leads between floes, and the presence or absence of brash or frazil ice (which can sorb the oil). Brash ice is the debris created when larger ice features interact and degrade. Frazil ice is the "soupy" mixture of very small ice particles that forms as seawater freezes. Slush ice is formed when snow settles on open water.

The key to the success of an individual burn in a broken ice field is, in part, controlled by how well the oil is contained by the ice it is in contact with. Other factors include oil weathering processes (i.e., evaporation and emulsification) and mixing energy from waves. Field experience has shown that it is the small ice pieces (i.e., the brash and frazil, or slush, ice) that will accumulate with the oil against the edges of larger ice features (floes) and control the concentration (i.e., thickness) of oil in a given area, and the rate at which the oil subsequently thins and spreads.

## 4.2. OIL ON WATER AMONG PACK ICE

In pack ice conditions the use of *in-situ* burning is controlled, to a large degree, by the concentration and types of ice present. In general, the applicability of burning can be divided into three broad ice concentration ranges:

- Open water and ice up to 3 tenths
- Between 3 and 7 tenths and
- Greater than 7 tenths.

In ice concentrations greater than 7/tenths, the ice will effectively contain the oil; if slicks are thick enough they can be burned effectively without additional containment (SL Ross and DF Dickins, 1987). In the lowest range, the oil's spread and movement will not be greatly affected by the presence of the ice, and open water *in-situ* burning techniques may be possible. This would generally involve the collection of slicks with fire boom operated by tow vessels, and their subsequent ignition. The ice concentration range from 3 to 7 tenths is the most difficult from an *in-situ* burning perspective. The ice will reduce the spreading and movement of the slick, but not sufficiently to allow burning without additional containment. The deployment and operation of booms in this ice concentration would be difficult, if not impossible. Untended booms could be deployed into the ice by helicopter, but the amount of oil that could be collected by this technique is unknown.

An important area of recent and ongoing research has been the use of chemical herding agents to thicken oil for burning. For oil slicks in ice concentrations of 1–7/tenths, the oil will likely spread out to an unburnable thickness, and it will be difficult or impossible to use containment booms to thicken the oil. The concept here is to apply a chemical herding agent to the water surrounding a thin slick; the herding agent causes the slick to contract and thicken such that burning may be possible. These were investigated for open water conditions in the 1970s, and small-scale tests in ice showed promise as long as the oil was fluid (i.e., above its pour point).

Over the last three years, testing has been done at increasing scales, culminating in an outdoor near-full scale test in Alaska in 2006. In general terms, the tests have involved spilling oil in a range of ice types and concentrations, allowing the oil to spread to an equilibrium (i.e., non-burnable) thickness, then applying herder to the perimeter of the oil slick. Within a few minutes, the herding agent caused the slick to contract to cover a much smaller area, and consequently, a much greater thickness. The technique appears to have considerable promise, and further full-scale field tests are planned for the future.

*In-situ* burning of oil spilled in pack ice during break-up will likely be easier than in the same ice concentration during freeze-up. In fall, the sea is constantly freezing, which generates significant amounts of slush ice which can severely hamper containment and thickening (naturally, or with booms) of slicks for burning; it is dark for much of the day, and it is cold, and only going to get colder with the onset of winter. During break-up, there is much less slush and brash ice present, the ice floes are deteriorating and melting, there is 24-h daylight and the temperatures are warming.

### 4.3.  OIL ON SOLID ICE

*In-situ* burning is the countermeasure of choice to remove oil pools on ice (created in the spring by vertical migration from an encapsulated oil layer or by drilling into an encapsulated oil lens in the ice sheet). There is a high degree of knowledge on the ignition and burning of oil on melt pools. For large areas of melt pools, helicopters deploying igniters would be used to ignite individual pools of oil. For smaller areas, manual ignition techniques could be employed.

Wind will generally blow oil on melt pools to the downwind ice edge, where it will be herded to thicknesses of approximately 10 mm. Individual melt pool burn efficiencies are thus on the order of 90%. The overall efficiency of *in-situ* burning techniques in removing oil from the ice surface ranges from 30% to 90%, with an average in the 60–70% range, depending on the circumstances of the spill (e.g., melt pool size distribution vs. igniter deployment accuracy, film thickness, degree of emulsification, timing of appearance vs. break-up, etc.). For areas where the oil surfaces early in the melt, it could be possible to manually flush and/or recover remaining burn residue.

Winds and currents will herd oil in leads to the downwind edge, where it can be ignited and burned. In leads where a current herds the oil against an edge, very high removal efficiencies can be obtained.

### 4.4.  OIL IN SNOW

In the case of oil spilled on the ice surface and mixed with snow, burning of oiled snow piles can be successfully achieved even in mid-winter Arctic conditions. Oiled snow with up to 70% snow by weight can be burned *in-situ*. For higher snow content mixtures (i.e., lower oil content), promoters, such as diesel fuel or fresh crude, can be used to initiate combustion. For the lower concentrations of oil in snow, the technique of ploughing oiled snow into concentrated piles may be the only way of achieving successful ignition and burning. In many cases, waiting for the snow to melt could result in thin oil films incapable of supporting combustion and spread over a large ice area. For this technique, the oiled snow is scraped into a volcano-shaped pile, with the centre of the "volcano" scraped down to the ice surface. A small amount of promoter is ignited in the centre of the pile. The heat from the flames melts the surrounding inside walls of the conical pile, releasing the oil from the snow that runs down into the centre and feeds the fire. This technique can generate considerable amounts of melt water, which needs to be managed.

## 5.  Technologies for Conducting *In-situ* Burns

A variety of methods are available to ignite an oil slick, including devices designed or modified specifically for ISB as well as simple, ad-hoc methods. Successful ignition of oil on water requires two elements:

- Heating the oil to its fire point, such that sufficient vapors are produced to support continuous combustion
- Providing an ignition source to start burning.

For light refined products, such as gasoline and some un-weathered crude oils, the flash point may be close to the ambient temperature and little if any pre-heating will be required to enable ignition. For other oil products, and particularly for those that have weathered and/or emulsified, the flash point will be much greater than the ambient temperature and substantial pre-heating will be required before the oil will ignite. The choice of one igniter over another for a given application will depend mainly on two factors:

- Degree of weathering or emulsification of the oil, which will dictate the required energy level of the igniter
- Size and distribution of the spill, which will determine the number of ignitions required to ensure an effective burn.

### 5.1. HELI-TORCH

The Heli-torch was originally developed as a tool for burning forest slash and for setting backfires during forest-fire control operations. It was adapted for use in ISB in the mid-1980s and found to be an effective system for igniting spilled oil. The Heli-torch has been tested extensively, used in a number of field trials, and refined considerably over the years, resulting in its being viewed as the igniter of choice for ISB.

The Heli-torch emits a stream of gelled fuel, typically gasoline that is ignited as it leaves the device. The burning fuel falls as a stream that breaks into individual globules before hitting the slick. The burning globules produce a flame that lasts for several minutes, heating the slick and then igniting it. The globules' burn time depends upon the fuel used and the mixing ratio of the fuel and gelling powder. Although gasoline is the fuel typically used, alternatives such as diesel, crude oil, or mixtures of the three fuels have been found to produce a greater heat flux, and they should be considered for highly weathered oils and emulsions that may be difficult to ignite.

## 5.2. HANDHELD IGNITERS

A variety of igniters have been developed for use as devices to be thrown by hand from a vessel or helicopter. These igniters have used a variety of fuels, including solid propellants, gelled kerosene cubes, reactive chemical compounds, and combinations of these. Burn temperatures for these devices range from 650°C to 2,500°C and burn times range from 30 s to 10 min. Most hand-held igniters have delay fuses that provide sufficient time to throw the igniter and to allow it and the slick to stabilize prior to ignition.

## 5.3. AD-HOC IGNITERS

For small, contained spills, simple ad-hoc techniques can be used to ignite the oil. For example, propane- or butane-fired weed burners have been used to ignite oil on water. As weed-burners or torches tend to blow the oil away from the flames, these techniques would only be applicable to thick contained slicks. Rags or sorbent pads soaked in fuel have also been successfully used to ignite small spills. Diesel is more effective than gasoline as a fuel to soak sorbents or rags because it burns more slowly and hence supplies more pre-heating to the oil.

Gelled fuel can also be used without the Heli-torch as an ad-hoc igniter. This was the method used for the test burn during the Exxon Valdez spill in 1989. Gasoline and gelling agent were mixed by hand in a plastic bag, and then the bag was ignited and allowed to drift into the slick contained within a fire-resistant containment boom.

## 5.4.  IGNITION PROMOTERS

Ignition promoters are used to increase the ignitability of an oil slick or to promote the spreading of flame over the surface of a slick. Petroleum products, such as gasoline, diesel, kerosene, aviation gasoline, and fresh crude oil, have all proved effective as ignition promoters. Of these, the middle distillates, such as diesel and kerosene, are preferred because they burn more slowly and produce a higher flame temperature. Crude oil is also very effective as it contains a mixture of com-ponents.

Emulsion-breaking chemicals can also be considered as ignition promoters. The concept is to apply the chemical to emulsified oil to break the emulsion *in-situ*, thus increasing the likelihood of successful ignition. Large-scale tests have proven the feasibility of this approach and research to include emulsion-breaking chemicals in the fuel of the Heli-torch system has been undertaken. There are presently no demulsifiers on the U.S. approved list of chemicals for oil spill use; however, their use as combustion promoters (which are permitted) is not specifically excluded. Most of the demulsifiers and other ignition promoters will be consumed in the resulting fire.

When using an ignition promoter, it is important to distribute the promoter over as large an area as possible. Simply pumping it onto one location of the slick will create a thick pool of the promoter in one area and it will not promote ignition effectively.

## 5.5.  FIRE-RESISTANT BOOM

To achieve an effective burn in lower ice concentrations, boom is required to create and maintain an oil thickness that will burn efficiently. The two main requirements for a fire-resistant boom are to provide oil containment (floatation, draft, and freeboard) and to resist fire damage. This section provides a brief description of the main types of fire-resistant boom. Additional detailed specifications are provided in the USCG Manual for products that are commercially available in the United States and that have been involved in recent fire-resistance testing.

Two main methods of providing fire resistance are used. Passive or intrinsically fire-resistant boom uses fire-resistant materials such as ceramic fibres or stainless steel. The active method keeps the boom materials within an acceptable range of temperatures by supplying coolant (usually water) to surfaces of the boom. Other ad-hoc methods of containment are also described at the end of this section.

A number of booms have been tested at the Oil and Hazardous Materials Simulated Environmental Test Tank (Ohmsett) and have been found to have similar containment limits as conventional boom, with first-loss tow speeds in the range of 0.85–1.0 knots when towed in calm water in a U-shape. Due to the weight of materials used for fire-resistance, the weight per unit length is generally much higher and the buoyancy-to-weight (b/w) ratio is much lower than for conventional booms of a similar size. Their lower b/w ratios mean that they are generally not applicable for high sea states. Fire-resistant booms often require special handling, partly due to their higher weight and due to the use of materials that are less rugged than those used in conventional booms.

Tests to confirm fire resistance have been performed in recent years and the American Society for Testing and Materials (ASTM) International has developed a standardized test (F 2152). The test comprises three 1-h burn cycles separated by two 1-h cool-down cycles during which the boom is exposed to waves. The test is designed to simulate the stresses that a boom would receive in a typical burn scenario, where the boom is used alternately to collect oil and then contain it during a burn. A heat exposure is specified to simulate the effects of a crude oil fire; in the test either burning diesel or using a specially designed propane system that is available at Ohmsett and provides an equivalent heat can supply the specified heat. Booms are judged to have passed the test if they survive and can contain oil at the conclusion of the cyclic heat exposure.

Based on these tests, there is recognition that many fire-resistant booms have a limited life when exposed to fire, which means that an extensive ISB operation will require the periodic replacement of boom, depending on the intensity and duration of the burn.

## References

Bech, C., Sveum, P. and Buist, I., 1993, The Effect of Wind, Ice and Waves on the *In-Situ* Burning of Emulsions and Aged Oils, in: *Proceedings of the Sixteenth Arctic and Marine Oilspill Program Technical Seminar.* Environment Canada, Ottawa, Ontario, Canada, pp. 73 5–748.

Brown, H.M. and Goodman, R.H., 1986, In Situ Burning of Oil in Ice Leads, in: *Proceedings of the Ninth Arctic and Marine Oilspill Program Technical Seminar.* Environment Canada, Ottawa, Ontario, Canada, pp. 245–256.

Buist, I., McCourt, J., Potter, S., Ross, S. and Trudel, K., 1999, In Situ Burning. *Pure Appl. Chem.* **71**(1) 43–65.

Buist, I., Coe, T., Jensen, D., Potter, S., Anderson, L., Bitting, K. and Hansen, K., 2002, *In-Situ Burn Operations Manual.* U.S. Coast Guard Research and Development Center, Groton, CT.

Buist, I.A., Ross, S.L., Trudel, B.K., Taylor, E., Campbell, T.G., Westphal, P.A., Myers, M.R., Ronzio, G.S., Allen, A.A. and Nordvik, A.B., 1994, *The Science, Technology and Effects of Controlled Burning of Oil Spills at Sea.* Technical Report Series 94-013, Marine Spill Response Corporation, Washington, DC.

Guénette, C.C., 1997, *In-Situ* Burning: An Alternative Approach to Oil Spill Clean-up in Arctic Waters, in: *Proceedings of the Seventh (1977) International Offshore and Polar Engineering Conference*, pp. 587–593.

Guénette, C.C. and Wighus, R., 1996, In Situ Burning of Crude Oil and Emulsions in Broken Ice, in: *Proceedings of the 19th AMOP Technical Seminar.* Environment Canada, pp. 895–906.

Shell Oil Company, Sohio Alaska Petroleum Company, Exxon Company, USA, Amoco Production Company, 1983, *Oil Spill Response in the Arctic – Part 2: Field Demonstrations in Broken Ice.* Shell Oil Company, Sohio Alaska Petroleum Company, Exxon Company, USA, Amoco Production Company. Anchorage, Alaska, August 1983.

S.L. Ross Environmental Research Ltd, 1983, *Evaluation of Industry's Oil Spill Countermeasures Capability in Broken Ice Conditions in the Alaskan Beaufort Sea.* Report by S.L. Ross Environmental Research Limited for Alaska Department of Environmental Conservation. ADEC, Anchorage, AK.

SL Ross Environmental Research Ltd. and DF Dickins Associates Ltd, 1987, *Field Research Spills to Investigate the Physical and Chemical Fate of Oil in Pack Ice.* Environmental Studies Research Funds Report no. 62. ESRF. Calgary.

SL Ross Environmental Research Ltd. and DF Dickins Associates Ltd., 2003, *Tests To Determine The Limits To In-situ Burning Of Thin Oil Slicks In Broken Ice.* Report to MMS and ExxonMobil Upstream Research. Herndon, VA.

Smith, N.K. and Diaz, A., 1987, In-Place Burning of Crude Oils In Broken Ice, in: *Proceedings of the 1987 Oil Spill Conference.* American Petroleum Institute, Washington, DC, pp 383–387.

Walton, W.D. and Mullin, J.V., 2003, *In-situ Burning of Oil Spills: Resource Collection.* National Institute of Standards and Technology. Gaithersburg, MD.

# RECENT MID-SCALE RESEARCH ON USING OIL HERDING SURFACTANTS TO THICKEN OIL SLICKS IN PACK ICE FOR *IN-SITU* BURNING

I. BUIST[†] & S. POTTER
*SL Ross Environmental Research Ltd., 200-717 Belfast Rd., Ottawa, Ontario, K1G 0Z4, Canada*

L. ZABILANSKY
*USACE Cold Regions Research and Engineering Laboratory, 72 Lyme Road, Hanover, New Hampshire, USA*

A. GUARINO
*Ohmsett Facility, P.O. Box 473, 381 Elden Street, MS 4021, Atlantic Highlands, NJ 07716, USA*

J. MULLIN
*U.S. Minerals Management Service, Engineering and Research Branch, 381 Elden Street, Mail Stop-4021, Herndon, VI 20170-4817, USA*

**Abstract.** Preliminary and small-scale laboratory testing at the scale of 1 and 10 $m^2$ of the concept of using chemical herding agents to thicken oil slicks among loose pack ice for the purpose of *in-situ* burning was completed in 2004. The encouraging results obtained from these tests prompted further research to be carried out. This paper will present the results of additional testing at larger scales at CRREL and at Ohmsett.

The additional phases of the work involved:

1. Conducting a test program at the scale of 100 $m^2$ in the Ice Engineering Research Facility Test Basin at the US Army Cold Regions Research and Engineering Laboratory (CRREL) in November 2005.

---

[†] To whom correspondence should be addressed. E-mail: ian@slross.com

W. F. Davidson, K. Lee and A. Cogswell (eds.), *Oil Spill Response: A Global Perspective.*   41
© Springer Science + Business Media B.V. 2008

2. Conducting a test program at the scale of 1,000 m$^2$ at Ohmsett in natural or artificial pack ice in February 2006.

A series of burn tests at the scale of 50 m$^2$ with herders and crude oil in a pit containing broken sea ice is planned for November 2006 in Prudhoe Bay, AK. The results of the first two phases of the testing will be presented and the plans for the November burn tests will be discussed.

**Keywords:** herding, oil slicks, pack ice, *in situ* burning, oil spill response

## 1.  Introduction

*In-situ* burning may be one of the few viable options to quickly remove oil spilled in loose pack ice. One fundamental problem for burning blowout slicks or subsea pipeline leaks in pack ice less than 6 to 7 tenths coverage, is that the slicks can either initially be too thin, or they can thin quickly. If these slicks could be thickened to the 2- to 5-mm range, effective burns could be carried out (SL Ross, 2003).

The use of surface-active agents, sometimes called oil herders or oil collecting agents, to clear and contain oil slicks on a water surface is well known. When applied on water, these agents have the ability to spread rapidly into a monomolecular layer, as a result of their high spreading pressure. Consequently, small quantities of these surfactants will quickly clear thin films of oil from large areas of water surface. The oil is contracted into thicker slicks. For application of herders in loose pack ice, the intention would be to herd freely-drifting oil slicks to burnable thickness, then ignite them from the air with a Helitorch. Burning would be performed without additional mechanical containment.

This paper describes the first two of three planned mid-scale tests carried out to explore the potential effectiveness of oil-herding agents in pack ice conditions.

## 2.  Background

Field deployment tests of booms and skimmers in broken ice conditions in the Alaskan Beaufort Sea highlighted the severe limitations of conventional equipment in even trace concentrations of broken ice (Bronson *et al.*, 2002). *In-situ* burning may be one of the few viable options to quickly remove oil spilled in such conditions.

The use of specific chemical surface-active agents, sometimes called oil herders or oil collecting agents, to clear and contain oil slicks on an open water surface is well known (Garrett and Barger, 1972; Rijkwaterstaat, 1974; Pope *et al.*, 1985; MSRC, 1995). These agents have the ability to spread rapidly over a water surface into a mono-molecular layer, as a result of their high spreading coefficients, or spreading pressures. The best agents have spreading pressures in the 40–50 mN/m range, whereas most crude oils have spreading pressures in the 10 to 20 mN/m range. Consequently, small quantities of these surfactants (about 5 L/km) will quickly clear thin films of oil from large areas of water surface, contracting it into thicker slicks.

Although commercialized in the 1970s, herders were not used offshore because they only worked in calm conditions: conventional containment booms are still needed in wind above 4 knots, and break-ing waves disrupt the herder layer. For application in loose pack ice, the intention would be to herd freely-drifting oil slicks to a burnable thickness, then ignite them with a Helitorch. The herders will work in conjunction with the limited containment provided by the ice to allow a longer window of opportunity for burning.

A very small scale (1 m$^2$) preliminary assessment of a shoreline-cleaning agent with oil herding properties was carried out to assess its ability to herd oil on cold water and among ice (SL Ross, 2004). The results were promising:

1. Using the shoreline cleaner on cold water (2°C) greatly reduced the area of sheens of fluid oils, but the thickness of the herded oil was only in the 1-mm range.

2. On thicker (ca. 1 mm) slicks, the shoreline cleaner effect was much more promising and could herd slicks to thicknesses of 2–4 mm.

3. Although the presence of ice forms in the pans slightly retarded the effectiveness of the herding agent, it still considerably thick-ened oil among ice.

4. The composition of the oil appeared to play a strong role in deter-mining potential efficacy: gelled oils that did not spread on cold water could not be herded.

Further tests were carried out to explore the relative effectiveness of three oil-herding agents in simulated ice conditions; conduct larger scale (10 m$^2$) quiescent pan tests to explore scaling effects; carry out small-scale (2–6 m$^2$) wind/wave tank testing to investigate wind and wave effects on herding efficiency; and, perform small-scale *in-situ* ignition and burn testing (SL Ross, 2005). The results from these experiments showed that the application of a herder to thin oil slicks in pack ice has considerable promise for thickening them for *in-situ* burning. One herder formulation proved to be the best suited for the cold conditions. The herded thickness produced by this formulation was consistently in the 3+ mm range for 1-L and greater slicks. Crude oil slicks herded by the chemical were successfully ignited and burned. The burn efficiencies measured were similar to those for physically contained slicks of the same dimensions. The encouraging results obtained from this and the previous study indicated that further research was warranted at a larger scale with the herder and with oils that are fluid at freezing temperatures.

Concern has been expressed regarding the potential toxicity risk to marine species of using herding agents in broken ice. These agents should not cause harm to the marine environment because they are of low toxicity and extremely small quantities are used. The toxicity data on the U.S. National Contingency Plan (NCP) web site indicates that EC 9580 is only about half as toxic as approved chemical dispersants and much less toxic than the oil itself. EC9580, and the main surface-active ingredients of many successful herders are not soluble in water (they are dispersible) and are not intended to enter the water column, only to float on the surface. Since herders are intended to form a monomolecular layer, the products are employed at very low application rates (5 L/km of spill perimeter, or $5 \times 10^{-2}$ g/m$^2$ = 0.05 US gal/acre of water surface) compared with dispersants (5 US gal/acre = 4.7 g/m$^2$) and, if dispersed, would produce concentrations in the water column far below levels of concern (dispersing the entire $5 \times 10^{-2}$ g/m$^2$ layer of herder into the upper metre of the water column would only produce a concentration of 0.05 ppm).

In light of the paucity of other viable, high encounter rate oil spill cleanup techniques for broken ice, further testing on the use of herders to enhance the potential for *in-situ* burning was undertaken. A workshop on Advancing Oil Spill Research in Ice-covered Waters sponsored by the United States Arctic Research Commission and the Prince William Sound Oil Spill Recovery Institute included this idea as one of their recommended program areas (Dickins, 2004).

The concept of pre-treating the water surface to prevent spills from rapidly spreading to unignitable thicknesses also deserved further research. Field tests of herders on open water with a 25-gal fuel oil slick in Chesapeake Bay (Garrett and Barger, 1972) and a 5-t crude oil slick in the North Sea (Rijkwaterstaat, 1974) have shown them to retain their effectiveness for several hours in winds of 6 m/s (12 knots) with 2-m (6-ft) seas. Restraining a slick on water from spreading for many hours among dynamic broken ice should be achievable and would offer a valuable extension in the window of opportunity for slick ignition.

One of the herder formulations tested proved capable of herding slicks that were fluid at ambient temperature among ice to 3–4 mm. This would allow ignition using conventional gelled gasoline igniters and result in 66–75% removal efficiencies (SL Ross, 2003). In a real spill situation, once a large, 3–4 mm slick of oil on water had been ignited around its periphery, it is likely that the inward air flow gene-rated by the combustion would further herd the oil to thicknesses of 10 mm (Buist, 1987), resulting in even higher oil removal efficiencies.

In November and December of 2005 a two-week test program was carried out at CRREL in New Hampshire using their indoor Ice Engi-neering Test Facility. A total of 17 individual tests were carried out in various concentrations of broken ice at a size scale of 81 $m^2$. In February 2006 a series of five tests was carried out at Ohmsett to explore the use of herders on spreading oil slicks in free-drifting ice fields at a scale of 1,000 $m^2$.

## 3. Testing at CRREL

The first series of mid-scale experiments was conducted in a large, refrigerated ice tank located at the US Army CRREL Ice Engineering Research Facility Test Basin in Hanover, NH. The main features of this facility (Figure 1) are:

- Basin dimensions of 37 m long × 9 m wide × 2.4 m deep.
- The basin is in a large refrigerated room with temperature control down to 24°C.

*Figure 1.* CRREL Ice Engineering Research Facility test basin.

- Ice sheets can be grown with a practical range of ice thickness from 2 to 15 cm, with the capability to grow and test multiple ice sheets each week.
- Two towing carriages and dedicated instrumentation and data acquisition systems.

For these tests, low-volatility petroleum oil (Hydrocal 300, a dearomatized lube stock oil with a nominal viscosity of 200 mPas and density of 0.88 g/cm$^3$ at 25°C – one of the test oils employed at Ohmsett) was used to eliminate potential problems with vapors in the CRREL facility. Once an ice sheet had been grown in the basin, it was divided into two 9 × 9 m sections using specially designed small oil booms and the target coverage of ice was created inside each area (Figure 2). Then a pre-measured volume of oil (25, 40 or 56 L for 70%, 50% or 30% ice cover) was poured onto the water surface between the floes using a spill plate to prevent the oil from getting under the ice (Figure 3). A video camera (inside an insulated cover) mounted above the centre of each test area was used to obtain overhead images

*Figure 2.* Test basin layout.

*Figure 3.* Oil being poured onto spill plate in test area.

of each test. The cameras were fitted with a fisheye lens to cover the entire test area. An image was obtained from the video signal by a computer and Web-posted every 15 s, and a VHS copy of the entire test was made as a backup. The digital images from the video (Figure 4)

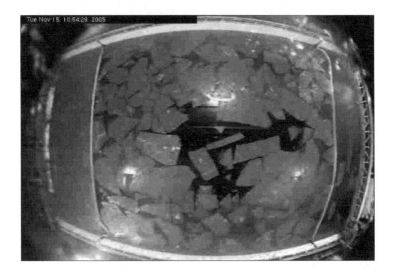

*Figure 4.* Image direct from overhead video.

*Figure 5.* Image after fisheye and horizontal corrections.

were corrected in PaintShop Pro® (PSP) using two transformations: the first used a plug in called PTLens to correct the fisheye distortion (Figure 5); the second used PSP's horizontal perspective correction. Next, the oil slick in the image was defined as black and everything else as white (Figure 6). Then, the image analysis software called Scion Image© was used to count the number of black pixels in each image. Finally, the pixel count was converted to area using scaling factors obtained from images taken of the test areas with known dimensions.

Once the oil had stopped spreading among the ice and a digital video image had been captured, the herding agent was applied around the edge of the slick at the recommended dose using a 3-mL syringe (Figure 7). Video images were captured for a period of 1 h after herder application. The images taken at nominally 1, 2, 5, 10, 20, 40 and 60 min after herder application were analyzed for oil slick area, which was converted to slick thickness using the measured volume of oil employed for the test. Duplicate tests and duplicate image analysis indicate that the error in the estimated thickness is likely within ±7.5%.

The test variables included:

- Ice coverage (10%, 30%, 50% and 70% surface coverage)
- Ice type (brash vs. frazil)
- Air temperature (0° vs. −21°C)
- Herder application time (post-spill vs. pre-spill) and
- Waves (calm vs. small waves)

*Figure 6.* Image with oil slick converted to black for area analysis.

*Figure 7.* Applying herder around periphery of slick.

In total, 17 tests were conducted over a two-week period. Following each series of two tests: the booms were removed sequentially; the ice from the test areas was pushed into the melt pit at the one end of the tank; the open water was cleaned of oil and herder with sorbent sweeps; the booms were cleaned with sorbent pads; and the test areas prepared for the next two tests. The cleanliness of the water in the test areas was confirmed by conducting an oil-spreading test with a small volume of oil inside a small floating ring placed on the water inside the test area.

Figure 8 shows the effect of the herder on the Hydrocal slicks in brash ice of different concentrations. Within the estimated error in thickness measurements, there is no difference in the effectiveness of the herder in 50% and 70% brash ice cover (Figure 5 shows a test in 70% ice cover); the oil spread to an equilibrium thickness of 3 mm (in 50% ice) to 4 mm (in 70% ice) and was then herded to 6–7 mm. The herded thickness declined slowly over the 1-h test.

Figure 9 illustrates the effect of ice type on the herding action. There appears to be no difference between the effects of the herder in 10% brash or frazil ice. In 50% frazil ice the oil did not spread initially to less than approximately 8 mm. Note that although the ice concentration was supposed to be 50% for this frazil test, the overhead images indicate much higher ice coverage, probably 90% or more composed of

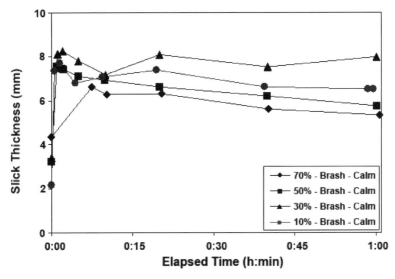

*Figure 8.* Herded slick thickness in various ice covers at the CRREL basin (brash ice, calm conditions, 0°C air).

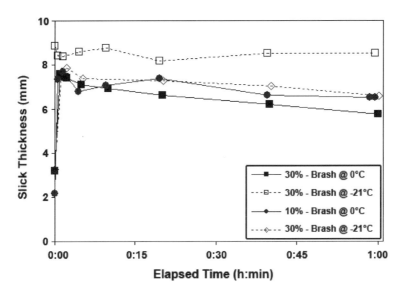

*Figure 9.* Comparison of herded thickness in frazil and brash ice at CRREL (calm conditions, air temperature = 0 except 50% frazil @ –7°C.

small crystals. Figure 10 demonstrates that the herder seems to work as well at air temperatures of –21°C as it does at 0°C.

Figure 11 shows that low wave action (with a 3-s period and a height of about 3 cm) did not significantly affect the herders action in the lowest ice concentration; however, in the 30%, 50% and 70% ice cover, the wave action and its effects on the ice field broke the slick into many small slicklets. In the 30% ice cover with waves, the herded slick remained as fairly large contiguous slicks for between 20 and 40 min whereas the same test in calm conditions resulted in large contiguous slicks after an hour. In the 50% ice cover in waves the slick remained contiguous for between 10 and 20 min. In 70% ice cover (with waves with a shorter period of 1 s) the waves quickly converged the ice into 90+% coverage that compressed the oil into small interstices among the ice.

Figure 12 shows very little difference in herded slick thickness if the herder was placed on the water before or after the oil, except in the lowest (10%) ice cover where pre-spill application of the herder resulted in significantly thicker slicks.

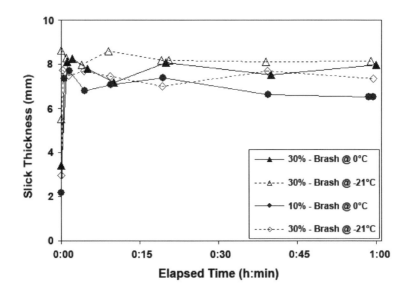

*Figure 10.* Effect of air temperature on herded slick thickness at the CRREL basin (brash ice, calm conditions).

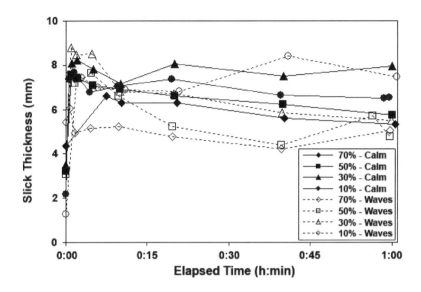

*Figure 11.* Effects of wave action on herded slick thickness at CRREL basin.

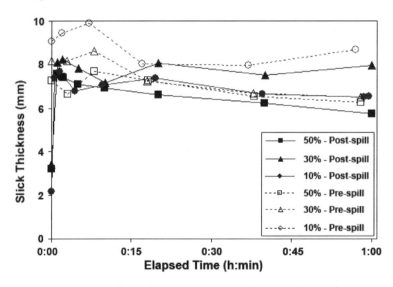

*Figure 12.* Comparison of applying herder to water before and after spilling oil (brash ice, calm conditions, 0°C).

## 4.  Testing at Ohmsett

The second series of mid-scale experiments was conducted at Ohmsett in Leonardo, NJ in February 2006. The purpose of these tests was to conduct experiments with herders at the scale of 1,000 m$^2$ using free-drifting slicks and ice pieces. The middle portion of the tank was divided into two 20 × 50 m test areas using small containment booms attached to the sides (Figure 13). The dividing booms were sealed tightly to the tank walls using clamped boom slides to allow them to move with waves. The ice was supplied by CRREL in the form of 1 × 1 × 20 cm slabs grown from urea-doped water to simulate sea ice. Each test involved placing 40 slabs in the test area (Figure 14), with 10 of the slabs quartered with an axe to provide a range of ice sizes. The tank water was maintained below 0°C using a large industrial chiller. The test oil was a 50:50 blend of Ewing Bank (26°API) and Arab Medium (30 °API) crude oils. For two tests evaporated crude was used. A drum of the crude was evaporated by bubbling compressed air until it had lost 11.3% by weight (representing 6-h exposure as a 1-mm slick). The volume of oil used in most tests was 60 L.

*Figure 13.* Test set-up at Ohmsett.

*Figure 14.* Adding ice slabs to test area.

Originally, it had been intended that the ice pieces would be placed inside a containment ring, and then the oil would be spilled into the ring and allowed to spread to equilibrium. Next, the ring would be lifted to release the oil and ice to spread and drift across or down the test area. This procedure was used for Test 1: however; it was apparent that, once the containment ring was lifted, the oil and ice drifted at very different velocities. It is believed that this was due to two factors. First, the fetch in the Ohmsett tank is quite small, and it is unlikely that the surface current generated by the prevailing wind extended more than a few millimetres below the surface of the water. This was enough for the oil to move with the induced surface current at the usual 3% of the wind speed, but not to move the ice pieces as quickly, with their much deeper draft. Second, the ice pieces (weighing upwards of 200 kg) required more time to accelerate than the oil slick. This problem was addressed by using two initial containment systems: a section of boom was used to contain the ice pieces just down-drift of the ring that held the oil at a thickness of approximately 3 mm (Figure 15). First, the boom holding the ice pieces was released, allowing the ice to drift. Once the ice was determined to be at full speed in the prevailing wind, the oil was released to drift into the ice field (Figure 16). When the oil slick was in the ice field, the herder was applied by two persons from the sides of the tank and from the bridges around the periphery of the test area using hand-held spray bottles.

*Figure 15.* Adding crude oil to containment ring.

*Figure 16.* Containment ring lifted to release oil to drift into ice.

The nominal dosage of herder was 50 g on the 100-m$^2$ test area. The test ended when the slick or ice reached a side or end or the test area. After two tests had been completed, the downwind containment booms were removed to allow the ice and oil to drift out of the area. Then fire monitors were used to herd any remaining oil and disperse any surfactant. Prior to each test, the surface of the test area was swept clean *with a sorbent* sweep.

A portable lift was used as a platform to take overhead pictures of the slick with a hand-held digital camera. The basket of the lift was raised to a height of 12.5 m above the water for each test. Photos of the test were taken at various times before and after the application of the herder for oil slick area analysis. The camera frame could not cover the entire slick area in some cases, and a series of overlapping shots were taken by moving the lift basket horizontally. These were digitally overlaid to form a composite photo (Figure 17). The same photo analysis technique used at CRREL was used to determine slick area. For each test, a few ice slabs were numbered, measured and used to scale the photos. At the present time, only an average scale has been applied to the test photos to estimate oil slick areas. Due to the additional inaccuracy introduced by the technique used at Ohmsett, the error in the estimated slick areas is likely higher than at CRREL, on the order of ±10% (compared to ±7.5%).

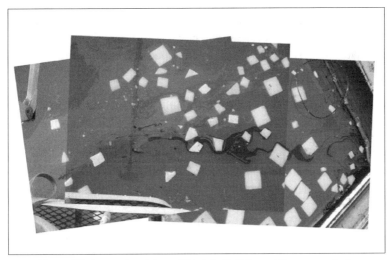

*Figure 17.* Composite picture of Test 5 oil slick used for area analysis.

I. BUIST ET AL.

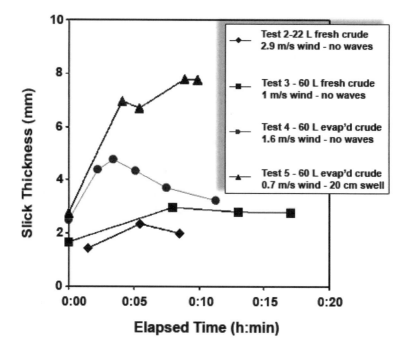

*Figure 18.* Ohmsett test results of herder in pack ice.

Figure 18 shows the results of all the successfully completed tests. In Test 2, 22 L of fresh crude was released from the containment ring 2 min after the ice was released. The herder application started 2 min after that. The first composite photo (and hence data point on Figure 18) was taken midway through the herder application (the initial photograph did not work). The second and third photo composites were taken 4 and 7 min later. In the time span between the first and second sets of photos the slick, though herded, began to break up into small slicklets under the influence of the 2.9 m/s wind. This behaviour may be related to the freshness of the crude (and hence its low viscosity) and the small volume of oil used for the tests (22 L on 1,000 m² of water surface). The slick was herded to an average thickness of approximately 2 mm over the 8½-min test.

In Test 3, the volume of fresh oil was increased to 60 L. The herder application commenced about 7 min after the oil was released from the containment ring. The test ended 11 min after the end of the herder application, when the slick reached a tank wall. The herder contracted the slick and maintained a slick thickness of 3 mm over the time of the test. With the greater oil volume (and perhaps the lower wind speed) the slick did not break into as many small slicklets as happened in Test 2; rather, it elongated into several "streamers" which resulted from the herder contracting individual "arms" of the initial slick (Figures 19 and 20).

*Figure 19.* Test 3, just prior to herder application.

*Figure 20.* Test 3, 11 min after herder application.

Test 4 involved the release of 60 L of evaporated crude. Herder application commenced 2½ min after the oil was released and was completed about 5 min later. The test ended about 6 min after the herder application finished, when the slick reached a tank wall. The herder initially contracted the slick to a thickness of more than 4 mm, but then streamers began to form as the slick drifted and the average thickness declined to 3 mm by the end of the test.

Test 5 entailed releasing 60 L of evaporated crude into the ice, allowing it to spread, then turning the wave generator at a low setting (9"-stroke and 10 cpm) to generate a 20-cm high swell with a 7-second period. The herder was applied after the waves had started, approximately 3½ min after the oil was released from the containment ring, and ending 4 min later. The test ended 7 min later. The herder contracted the slick to a thickness of 7 mm, and maintained it throughout the test period. Perhaps the wave action distributed and maintained the mono-molecular layer of herder better than in calm conditions. Figure 21 shows the slick just prior to herder application, for comparison with Figure 17, which shows the slick at the end of the test.

*Figure 21.* Test 5, prior to herder application (compare with Figure 17).

## 5. Future Plans

The next step in the research program will be performed at Prudhoe Bay, AK in the fall of 2006. A series of burn tests will be performed at a scale of 50 m$^2$ with herders and crude oil in a pit containing broken sea ice. The tests will be conducted in a shallow, lined pit. The dimensions of the pit will be about $8 \times 8 \times 20$ cm deep. Pieces of broken ice grown in an adjacent pit will be placed in the test area to create different concentrations of broken ice for the tests. The pit area will be completely covered with 1 mm of oil using 64 L; assuming the oil is herded to 3 mm thick this would equate to a circle with a diameter of 5.2 m providing a full-scale test fire. Gelled gasoline will be used as the igniter. Lengths of disused fire boom will be used to protect the edges of the pit. The effects of the herding agent will be quantified by measuring the change in surface area of a slick after treatment using overhead video and digital photography. Ignition and burn parameters will be observed, recorded on video, and determined by weighing the burn residue collected after each test.

After completing the burn tests, a final report describing all three mid-scale tests will be prepared and submitted to the project sponsors.

## 6. Summary

Two of the three planned test phases for this series of experiments on the use of chemical herders in pack ice have been completed. Although conclusions cannot be drawn at this point, the results, as analyzed to date, show that there is still considerable promise for the application of chemical herders to contract oil slicks in pack ice to thicknesses conducive to efficient *in situ* burning, particularly in light wind conditions. One more series of burn tests is planned, as is additional analysis of the experimental results.

## 7. Acknowledgements and Disclaimer

The Minerals Management Service (MMS), ExxonMobil Upstream Research Company and its Petroleum Environmental Research Forum (PERF) partners funded the research described in this paper. Dr. Tim Nedwed of ExxonMobil URC oversaw the project for PERF. Mr Joe Mullin of MMS was the COTR for MMS. ExxonMobil Upstream Research Company and MMS have reviewed this paper for technical adequacy according to contractual specifications. The opinions, conclusions, and recommendations contained in this report are those of the authors and do not necessarily reflect the views and policies of MMS or PERF. The mention of a trade name or any commercial product in this report does not constitute an endorsement or recommendation for use by MMS or PERF.

## References

Bronson, M., Thompson, E., McAdams, F. and McHale, J., 2002, Ice Effects on a Barge-Based Oil Spill Response Systems in the Alaskan Beaufort Sea, in: *Proceedings of the Twenty-fifth Arctic and Marine Oilspill Program Technical Seminar*, Environment Canada, Ottawa, Canada, pp. 1253–1269.

Buist, I.A., 1987, A Preliminary Feasibility Study of *In Situ* Burning of Spreading Oil Slicks", in: *Proceedings of the 1987 Oil Spill Conference*, American Petroleum Institute, Washington, DC, pp. 359–367.

Dickins, D., 2004, *Advancing Oil Spill Research in Ice-covered Waters*, Workshop sponsored by the United States Arctic Research Commission and the Prince William Sound Oil Spill Recovery Institute, PWSRCAC, Valdez, AK.

Garrett, W.D. and Barger, W.R., 1972, *Control and Confinement of Oil Pollution on Water with Monomolecular Surface Films*, Final Report to U.S. Coast Guard, November 1971, Project No. 724110.1/4.1 (also reprinted as U.S. Navy Naval Research Laboratory Memorandum Report 2451, June 1972, AD 744–943).

Marine Spill Response Corporation (MSRC), 1995, *Chemical Oil Spill Treating Agents*, MSRC Technical Report Series 93–105, Herndon, VA.

Pope, P., Allen, A. and Nelson, W.G., 1985, *Assessment of Three Surface Collecting Agents during Temperate and Arctic Conditions*, in: Proceedings of the 1985 Oil Spill Conference, API/EPA/USCG, Washington, DC, pp. 199–201.

Rijkwaterstaat, 1974, *Shell Herder Trials*, Report to the Dutch Ministry of Transport, Gravenhage, Holland.

SL Ross Environmental Research, 2003, *Tests To Determine The Limits To In Situ Burning of Thin Oil Slicks In Broken Ice*, Report to MMS and ExxonMobil Upstream Research, Houston, TX.

SL Ross Environmental Research, 2004, *Preliminary Research on Using Oil Herding Surfactant to Thicken Oil Slicks in Broken Ice Conditions*, Report to ExxonMobil Upstream Research, Houston, TX.

SL Ross Environmental Research, 2005, *Small-Scale Test Tank Research on Using Oil Herding Surfactants to Thicken Oil Slicks in Broken Ice for In Situ Burning*, Report to ExxonMobil Upstream Research, Houston, TX.

# WEATHERING OF OIL SPILLS UNDER ARCTIC CONDITIONS: FIELD EXPERIMENTS WITH DIFFERENT ICE CONDITIONS FOLLOWED BY *IN-SITU* BURNING

P.J. BRANDVIK[†]
*The University centre at Svalbard (UNIS),*
*Longyearbyen, Norway & SINTEF Materials and*
*Chemistry, Trondheim, N-7465, Norway*

L-G. FAKSNESS
*The University centre at Svalbard (UNIS),*
*Longyearbyen, Norway*

D. DICKINS
*DF Dickins Associates Ltd., California, USA*

J. BRADFORD
*Boise State University, Idaho, USA*

**Abstract[*].** The knowledge regarding weathering processes in Arctic oil spills and especially with the presence of ice is limited. Experimental studies have been performed in laboratories, but only to a limited degree in the field. This presentation summarized and compared results from field experiments performed in Norway in 1989, 1993 and 2003–2006. Two full-scale field measurements from experimental oil releases in Norway are initially used to compare the behavior of oil spilled in open water and in an Arctic broken ice scenario. Similar oil types and amounts (25–30 m$^3$) were used in an experimental oil release in open water at Haltenbanken (65° N) in 1989 and in dynamic broken ice at Svalbard (75° N) in 1993. Results from small-scale field experiments performed later (2003–2006) on Svalbard are also discussed and compared to the earlier field data. Several weathering properties for the oil spill in broken ice are strongly influenced by the low temperature,

---

[†] To whom correspondence should be addressed. E-mail: per.brandvik@sintef.no
[*] Full presentation available in PDF format on CD insert.

W. F. Davidson, K. Lee and A. Cogswell (eds.), *Oil Spill Response: A Global Perspective.*      63
© Springer Science + Business Media B.V. 2008

reduced oil spreading and reduced wave action caused by the high ice coverage. Reduced water uptake, viscosity, evaporation and pour point extend the operational time window for several contingency methods compared to oil spills in open waters. This could open up for dispersant treatment and *in-situ* burning even after an extended period of weathering for an oil spill in broken ice. In the period of 2003–2006, SINTEF and the University Centre in Svalbard (UNIS), and later also co-workers from US (DF Dickins and Boise State University) have performed field weathering studies with oil entrapped in and under ice followed with *in-situ* burning. This presentation will present recent results from these field experiments regarding oil weathering under different ice conditions (slush ice, 30% and 90% ice coverage) and the consequences for the ignitability and *in-situ* burning of the weathered oil. This research has been funded by Norwegian and US authorities (the Norwegian Research council and US Mineral Management Services) and several oil and industry companies (Statoil, Norsk Hydro, Shell, Alaska Clean Seas, ExxonMobile, ConnocoPhillips and Store Norske Spitsbergen kullkompani).

**Keywords:** oil spill response, arctic, ice, *in situ* burning

# DISPERSANT EFFECTIVENESS EXPERIMENTS CONDUCTED ON ALASKAN CRUDE OILS IN VERY COLD WATER AT THE OHMSETT FACILITY

J.V. MULLIN[†]

*U.S. Minerals Management Service, Engineering and Research Branch, 381 Elden Street, Mail Stop-4021, Herndon, VI 20170-4817, USA*

**Abstract[*].** In the winter of 2003, five Alaskan crude oils were tested in very cold water with Corexit 9527 dispersant at Ohmsett – The National Oil Spill Response Test Facility located in Leonardo, New Jersey. Ohmsett is a large outdoor, above ground concrete test tank that measures 203 m long by 20 m wide by 3.4 m deep. The tank is filled with 9.8 million litres of crystal clear salt water. At the south end of the tank there is a wave generator capable of producing waves 1 m in height and at the opposite end there is a moveable beach. The tank is spanned by a bridge system capable of towing full size oil spill response equipment at speeds up to 6.5 knots and is equipped to distribute test oils on the surface of the water at reproducible thicknesses. The National Academy of Science (NAS 2005) reviewed the test methods and results of the 2003 Alaskan Cold Water dispersant experiments (DE) and recommended that the U.S. Minerals Management Service (MMS) repeat the test program utilizing the improvements that have been made in the testing methods, protocols and instrumenttation at Ohmsett. In February–March 2006, MMS repeated the (DE) experiments in very cold water using four Alaskan crude oils (Alaska North Slope, Endicott, Northstar and Pt.McIntye) and Corexit 9527 dispersant. Oils were tested fresh, weathered by removal of light ends using air sparging and weathered by placing the oils in the tank in both breaking wave conditions and non-breaking waves. Results from these experiments will be presented that show in all DE tests Corexit 9527 dispersant was more than 90% effective in dispersing the crude oils tested in very cold water.

**Keywords:** wave tank, oil spill response, Ohmsett, dispersants

---

[†] To whom correspondence should be addressed. E-mail: Joseph.Mullin@mms.gov

[*] Full presentation available in PDF format on CD insert.

W. F. Davidson, K. Lee and A. Cogswell (eds.), *Oil Spill Response: A Global Perspective.* 65
© Springer Science + Business Media B.V. 2008

# ENHANCED CHEMICAL DISPERSION USING THE PROPELLER WASH FROM RESPONSE VESSELS

T. NEDWED[†] & W. SPRING
*ExxonMobil Upstream Research Company, P.O. Box 2189, Houston, TX 77252, USA*

**Abstract**[*]. Concentrated ice cover in a marine environment reduces the wave energy needed to disperse chemically-treated oil slicks. ExxonMobil is currently evaluating a concept to utilize the propeller wash from vessels to enhance chemical dispersion. Tests in an arctic basin utilizing a 1:25 scale model of an azimuthal-stern-drive platform-standby icebreaker resulted in effective dispersion (over 90% in most cases) on light oil and confirmed that such an icebreaker working in a sea-ice environment could efficiently contact and disperse oil. Future plans are to evaluate a diversion-boom system that will allow using propeller wash from conventional platform-standby or response vessels in both light-ice and open-water conditions.

**Keywords:** dispersant, propeller wash, azimuthal-stern-drive, oil spill response

---

[†] To whom correspondence should be addressed.  E-mail: Tim.j.nedwed@exxonmobil.com
[*] Full presentation available in PDF format on CD insert.

W. F. Davidson, K. Lee and A. Cogswell (eds.), *Oil Spill Response: A Global Perspective.*    67
© Springer Science + Business Media B.V. 2008

# EVALUATION OF ARCTIC *IN-SITU* OIL SPILL RESPONSE COUNTERMEASURES

K. LEE[†]

*Center for Offshore Oil and Gas Environmental Research, Bedford Institute of Oceanography, Fisheries and Oceans Canada, P.O. Box 1006, Dartmouth, Nova Scotia, B2Y 4A2, Canada*

**Abstract**[*]. With an anticipated increase in marine transport in the Arctic associated with global climate change and the expansion of Arctic offshore oil and gas exploration activities, the need for research on oil spill response technologies for use in ice covered waters has been identified as a priority. Two oil spill countermeasures based on the acceleration of natural recovery rates have been proposed for study in field trials within Arctic Canadian waters. These are: (1) the promotion of Oil Mineral Aggregate (OMA) formation; and (2) the use of chemical oil dispersants. Both techniques are based on the dispersion of oil from the surface into the water column; the premise being that resultant concentrations will be below the threshold limits that cause detrimental biological effects. Furthermore, oil disassociated in the form of micron-sized droplets or in association with mineral fines has an expanded oil-to-water surface area that results in enhanced microbial degradation. Hence, oil is effectively removed from the environment. In terms of use in the Arctic, both of these *in-situ* methodologies may offer a major operational advantage as there is no need for the physical removal and treatment of contaminated waste materials for treatment. This international project will provide fundamental scientific knowledge, field validation of response technologies, training to Arctic based oil spill responders; and build confidence and trust among stakeholders (public, industry and government).

**Keywords:** arctic, oil spill response, oil droplets, OMA

[†] To whom correspondence should be addressed. E-mail: LeeK@mar.dfo-mpo.gc.ca

[*] Full presentation available in PDF format on CD insert.

W. F. Davidson, K. Lee and A. Cogswell (eds.), *Oil Spill Response: A Global Perspective.*    69
© Springer Science + Business Media B.V. 2008

# OMA FORMATION IN ICE-COVERED BRACKISH WATERS: LARGE-SCALE EXPERIMENTS

D. CLOUTIER[†] & B. DOYON
*Fisheries and Oceans Canada, Hydraulic Engineering Division, 101 Champlain Boulevard, Mail Stop: QBC, Québec, G1K 7Y7, Canada*

**Abstract.** Investigation on oil-mineral aggregate (OMA) formation in ice-covered waters was conducted by the Canadian Coast Guard (CCG). Different tests were performed in large basins using Heidrun oil and calcite to determine the efficiency of the OMA formation in these conditions. The main objectives of the experiments were (1) to validate oil droplets (OD) and OMA formation in ice-covered waters for different turbulence levels, (2) to investigate the type and size of OMAs formed, and (3) to help identify performance indicators of the OMA process in these conditions. The results have shown that OD and OMA formation is initiated when the stirring action upsets the water surface in the basin. The majority of the OMAs formed were droplet OMAs and their sizes related to the energy level and to the ice type. The sur-face flow pattern with and without ice was characterised using a particle image velocimetry (PIV) method. The results have shown the circu-lation pattern of surface currents is modified due to the ice blocks. The energy damping and a significant reduction in current intensity were the most evident effects consequential to the presence of ice.

**Keywords:** OMA, ice, oil spill response, oil droplets

## 1. Introduction

The oil-mineral aggregation process is recognized as a highly positive mechanism in the self-cleansing of low energy shorelines (Bragg and Yang, 1995; Lee *et al.*, 1997). The OMA process is the interaction

---

[†] To whom correspondence should be addressed. E-mail: CloutierDA@dfo-mpo.gc.ca

W. F. Davidson, K. Lee and A. Cogswell (eds.), *Oil Spill Response: A Global Perspective.* 71
© Springer Science + Business Media B.V. 2008

between oil droplets and fine mineral in suspension in the water column that leads to the formation of small aggregates which have the particularity to limit the oil re-coalescence at the surface as well as the oil adhesion to the substrate, to be degraded by microorganisms and to be dispersed by currents (Weise *et al.*, 1997; Lee *et al.*, 2003). Oil-mineral aggregation occurs naturally on coasts where hydraulic energy produced by waves and currents is sufficient for oil droplet formation and interaction with suspended sediments (Cloutier *et al.*, 2003).

The aggregation process was used with success in the coastal zone affected by the Sea Empress in Wales in 1996. The clean up method consists of transferring the oiled material from the upper parts of the beach into the surf zone where wave action enhanced the OMA process and contributed to more than 50% of the oil dispersion in a just few days (Lee *et al.*, 1997). Surf washing was subsequently tested with success in a cold water environment characterised by limited wave energy (~30 cm) (Guénette *et al.*, 2003; Lee *et al.*, 2003; Sergy and Goodman, 2003). These results from the *Svalbard Shoreline Field Trials* showed that the oil-mineral aggregation process is efficient in cleaning sediments from low energy shorelines. Oil-mineral aggregation process with IF30 oil was investigated in the littoral zone (Lee *et al.*, 2003; Sergy and Goodman, 2003; Sergy *et al.*, 2003) which was characterized by cold water temperatures (3–7°C) during ice-free periods. The results of sediment relocation experiments have shown that, at both sites under investigation, the amount of oil remaining in the experimental plots was dramatically reduced within five days following sediment relocation treatments. Water sample analysis revealed that OMA formation occurred naturally on the oiled beaches at both sites and was accelerated by the sediment relocation procedure. The water sample analysis also demonstrated that a significant fraction of the oil dispersed into near shore waters and sediments was biodegraded. Although degradation rates were slowed in comparison to warmer water conditions, degradation followed the same pattern than that observed under warmer temperatures (Guénette *et al.*, 2003).

In the light of these results, the CCG has conducted a research program that aims to elaborate an alternate spill countermeasure using the OMA process in ice-covered waters. To date, no studies have been undertaken on the oil-mineral aggregation process as a potential countermeasure in the eventuality of an oil spill in ice-encumbered waters. Laboratory work has shown a certain efficiency of the oil-mineral aggregation process for different oil and sediment types in cold brackish waters (Khelifa *et al.*, 2002), and in the presence of ice (Lee, 2002; Khelifa and Lee, 2004). During winter months in the St. Lawrence River, the presence of an ice field and the lack of sufficient particles in the water column would likely affect the natural OMA process (Cloutier, 2004). As both these parameters are critical to the formation of OMAs, it was suggested that these limiting factors could be overcome by the introduction of sediments and mechanical energy into the environment. The use of boat propellers and fire hoses are proposed to produce turbulent flow and to introduce sediments in the water column respectively. This paper reports the results of a series of tests conducted in large basins at the CCG – Quebec Region, during winter 2006.

## 2.   Materials and Methods

All experiments were conducted in a large basin ($2.44 \times 2.44 \times 0.76$ m) made of a metallic frame covered by a plastic cloth (volumetric capacity = 4,500 *l*) and set up on the pier at the CCG, Quebec during winter 2006 (Figure 1). Prior to each experiment the basin was filled with brackish water (18%) up to 3,000 *l* with water temperature varying between

*Figure 1.* Top view of the basin filled with frozen brackish water. The ice cover is broken down before the experimentation.

0 and −2.0°C. Turbulent flow was produced in the water column by a boat propeller attached to a rail system fastened to the basin. The propeller was connected and activated by a generator (900–1,800 rpm).

Three cameras were used, one of which was an under water camera fixed on the rail system to document the OMA process and behaviour underneath the ice (−30 cm). Two aerial cameras were used to film the oil/water/ice surface movement and to document oil interaction with ice. One of the aerial cameras was used to characterise the flow turbulent characteristics with and without ice using a particle image velocimetry (PIV) method. Subsurface currents were characterised using an acoustic doppler velocimeter.

Tests were performed in the presence of slush ice and ice rubble. Slush ice was formed in the basin by brackish water freeze up during the night, while the ice rubble was formed either with fragments of the ice cover or commercial ice blocks. Ice characteristics such as thickness, quantity and porosity were measured before each test. The experiments were conducted with an 80% ice cover.

Prior to each experiment, 10 $l$ of Heidrun crude oil from Norway were poured onto the broken ice cover. The oil slick characteristics (thickness, shape and size) were noted before the activation of the boat propeller. Heidrun oil is quite often in transit on the St. Lawrence River since it is refined at Jean-Gaulin refinery of Saint-Romuald. Heidrun oil dynamic viscosity at 0°C is of 35 mPa·s and its surface tension is 18.4 mN/m. The latter remains relatively stable with temperature increase. Heidrun oil specific gravity is 0.85 kg/m$^3$.

The sediments used in the study are commercially obtained calcium carbonate. The mean grain size of the sediments is 0.03 mm which is classified as silt. The mineral composition of the sediments, expressed in percent weight, is 95–98 for calcium carbonate (calcite). Sediments and water were mixed in a tank prior to introduction onto the slick surface (Figure 2). Lee (2002) suggested that the procedure for sediment application may be determinant on the oil dispersion process and that the spreading of minerals by delivering a concentrated suspension would provide the best results. Therefore, sediments were delivered at the oil/ice/water interface as a concentrated mixture. During each test, the oil slick poured onto the ice surface was treated using a hosepipe connected to the tank (Figure 3). Since different ice types were used for the experimentation, a single concentration was used to test OMA formation (SSC = 1,000 mg/L).

*Figure 2.* Tank used for sediment/water mixture and introduction onto the oil slick.

*Figure 3.* The entire oil slick is treated with a sediment/water mixture introduced uniformly onto the surface of the slick.

Water samples were collected at regular intervals ($t = 10$ min) during each test. The samples were analysed via light microscopy and with photographs of the samples taken immediately after sampling to determine the type and size of aggregates formed under different hydraulic energy conditions and different ice types. Samples collected during the same interval were sent for total hydrocarbon content (THC) analysis.

## 3.  Results and Discussion

### 3.1.  HYDRODYNAMIC CONSIDERATIONS

Surface currents in the basin were measured using a PIV method for conditions with and without ice. For both cases, surface velocities were measured at 385 positions in images pulled from a 5-min recording at a frequency of 30 images/s. The camera resolution is $720 \times 480$ pixels so each pixel of the image corresponded to a $0.11$ m$^2$ surface of the basin. For the case involving ice, the surface water motion is assumed to be identical to the movement of the blocks. The main objectives of these measurements were to assess the damping effect of ice.

The results are presented as velocity vectors which give the current intensity and direction, and as velocity fields in Figure 4A and B, and in Figure 5A and B. The results show surface currents that are structured, as they form two contiguous circulation cells flowing in opposite directions (Figures 4A and 5A). This pattern is typical of flows generated in such environments. The comparison between the two cases shows that the ice acts as a perturbing force on the flow. The two contiguous cells are somewhat nonsymmetric due to the clogging action of the ice that piles up and is unevenly distributed. But the most noticeable effect due to the presence of the ice remains the reduction of the surface velocities, with values being reduced roughly by a factor of 20 (Figure 5B).

### 3.2.  OIL BEHAVIOUR IN THE PRESENCE OF ICE

The oil behaviour in the presence of slush and ice rubble was investigated. Ten (10) litres of oil were poured onto the ice surface before activation of the propellers at the lowest energy level. It was observed that the oil slick thickness and width vary with ice type. In the presence of ice rubble, the oil slick had a round shape with a diameter of around

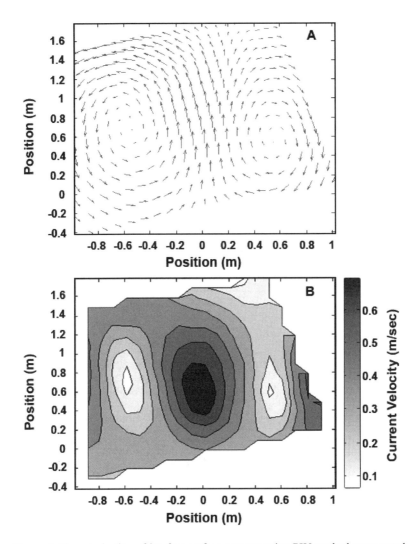

*Figure 4.* Characterization of ice-free surface currents using PIV method represented as vectors (A) and velocity fields (B).

2 m and a mean thickness of 0.4 cm (Figure 6A). It was also noted that when the oil was poured onto the ice rubble, oil droplets were forming at the slick margin (Figure 6B). The turbulence being at a very low level, this observation suggests that oil droplet formation with Heidrun crude oil in ice-covered waters could occur relatively easily.

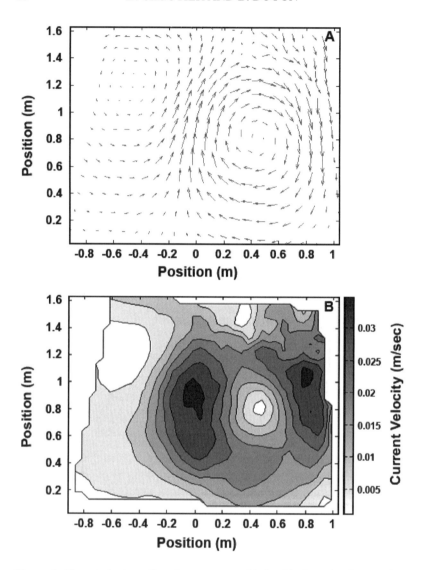

*Figure 5.* Characterisation of surface currents with ice blocks using PIV method represented as vectors (A) and velocity fields (B).

The slush ice cover was heterogeneous and thus slowed down the oil slick migration on the ice cover (Figure 6C). The oil spread in different directions, forming interdigitated structures with a total length

*Figure 6.* Oil slick appearances between ice rubble (A), oil droplets formation at the slick margin (B), and (C) oil migration pattern through slush ice.

inferior to previous observations (and thicknesses varying from 0.3 cm to more than 1.5 cm). The overall observations suggest that oil spilled in ice-covered waters will form a slick with both types of ice and that the slick extent and thickness will be closely related to the ice type.

After the oil was poured onto the ice surface, the stirring energy was introduced into the basin by regular increments of boat propeller revolution, from 900 to 1,800 rpm (900, 1,200, 1,500 and 1,800 rpm). Water samples collected at regular intervals ($t$ = 10 min) were analysed for oil droplet size. The results have shown that oil droplets form immediately when stirring energy is added to the water column independently of ice type. Basic visual observations have shown a constant decrease in OD size with the increase in turbulence level (Figure 7A–C). The OD sizes were measured for all samples collected.

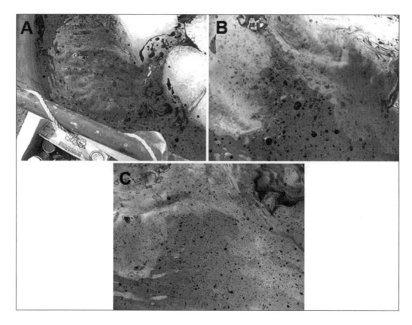

*Figure 7.* Oil droplet formation after activation of the propellers (A). The decrease in OD size with the increase of stirring energy level is obvious (B and C).

### 3.3.  OIL DROPLET SIZE ANALYSIS IN THE PRESENCE OF ICE AND DIFFERENT TURBULENT ENERGY LEVELS

The oil droplet size formed under different turbulent energy levels was analysed (1,500 and 1,800 rpm). The analysis of the OD size formed was achieved using a particle sizing technique (PST). The results show that in the presence of slush, oil droplets remained trapped in the ice matrix (Figure 8). Oil droplets formed in the presence of ice rubble were also visible in water samples but it is doubtful that they remain unaltered during or following sampling.

The analysis of the OD size formed in slush ice for two different stirring energy levels is presented in Figure 9. The general trend shows that oil droplet size ranges from 0.1 to 5.2 mm and that a larger proportion (64%) of smaller oil droplets (0.1 < OD < 0.5 mm) forms at the highest energy level. These results are consistent with previous observations (Delvigne *et al.*, 1987; Cloutier *et al.*, 2003).

It is also noted that for the lowest level of mixing energy (1,500 rpm), oil droplet sizes from 0 to 5.0 mm are measured, while at 1,800 rpm, oil droplet diameters remain smaller than 2.0 mm.

*Figure 8.* oil droplets trapped inside the slush ice matrix after sampling.

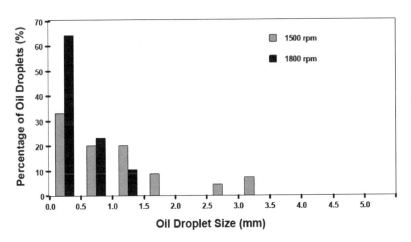

*Figure 9.* Oil droplet size variations for two different stirring energy levels.

These results suggest that in the presence of slush ice, a higher energy level is required to form a larger proportion of smaller oil droplets. As previously mentioned, slush ice acts as a trap with oil droplets. Thus, it is likely that oil droplets, once formed, remain trapped in the ice matrix.

### 3.4. ANALYSIS OF THE OMA TYPE AND SIZE FORMED IN THE PRESENCE OF DIFFERENT TYPES OF ICE

The analysis of the OMA type formed in the presence of slush ice and ice rubble reveals that nearly every aggregate observed was of droplet type. Photographs of droplet OMAs observed in the present set of experiments are presented in Figure 10A and B. Droplet OMAs consist of oil droplets covered by mineral particles on the external surface (Figure 10A). The aggregation of two or more droplet OMAs leads to the formation of multiple droplet aggregates (Figure 10B).

### 3.5. ANALYSIS OF THE OMA TYPE AND SIZE FORMED IN THE PRESENCE OF DIFFERENT TYPES OF ICE

The analysis of the OMA type formed in the presence of slush ice and ice rubble reveals that nearly every aggregate observed was of droplet type. Photographs of droplet OMAs observed in the present set of experiments are presented in Figure 10A and B. Droplet OMAs consist of an oil droplets covered by mineral particles on the external surface (Figure 10A). The aggregation of two or more droplet OMAs leads to the formation of multiple droplet aggregates (Figure 10B).

The analysis of the OMA size distribution was achieved using PST. Figure 11 provides a good idea of the number and size of the OMAs formed. It also demonstrates that a major proportion of the aggregates remain in suspension after sampling.

The results of the OMA size analysis is presented in Figure 12. The results show that the proportion of OMAs formed with slush ice and a SSC $\cong$ 1 g/L is smaller than 1.0 mm, the mean diameter being $\bar{x}$ = 530 µm (mean number of OMAs per sample = 98). These results clearly show that, independently of the mixing period, a large proportion of the OMAs formed are smaller than 1.0 mm. In fact, small OMAs (<500 µm) form the majority of the aggregates observed. For a

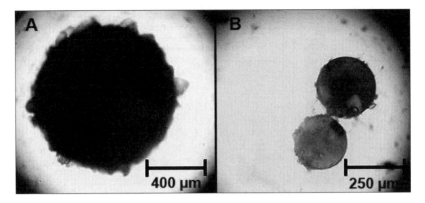

*Figure 10.* Figures of droplet OMAs formed in the presence of ice. Typical droplet OMA (A) consists of an oil droplet surrounded by mineral particles. Several droplets aggregated form multiple droplet OMAs (B).

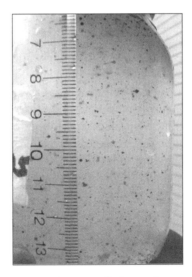

*Figure 11.* Water samples containing small OMAs in suspension, formed in the presence of ice rubble.

given level of energy, the OMA concentration increases with time. In particular, 42% of OMAs smaller than 500 μm are found after 10 min of mixing, 58% after 35 min of mixing, and 74% and 86% after periods of 59 and 75 min of mixing respectively. For OMAs of 500 μm and

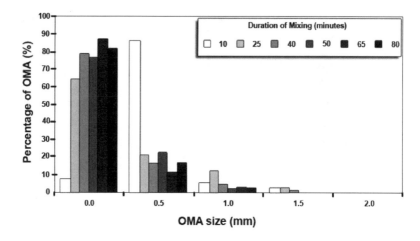

*Figure 12.* Distribution of the OMA size formed with slush ice.

more, the inverse situation seems to prevail. In fact, the longer the mixing period, the smaller the OMAs get. The OMAs having a size of 1.0 mm and more account for less than 15% (after 10 min of mixing) of the total OMA concentration, and continually decreases.

The same type of analysis was conducted for oil-mineral aggregates formed in the presence of ice rubble and similar sediment concentration. The results show a similar trend to that of the OMAs formed with slush ice, with the difference that a larger proportion of OMAs are <500 μm for all mixing times (Figure 13). The mean diameter is $\bar{x}$ = 470 μm (mean number of OMAs per sample = 68). Between 64% and 87% of the OMAs formed are found between 25 and 80 min of mixing respectively. This is consistent with laboratory results of Khelifa and Lee (2004) where OMAs were formed in cold water (without ice). They reported optimal mixing period of 20 min to form a significant proportion of OMAs with Heidrun oil (SSC = 200 mg/L).

A proportion of 86% of aggregate sizes of 0.5 < S < 1.0 mm are observed after 10 min of mixing. For mixing periods greater than 10 min, the proportion of OMAs found larger than 500 μm remains small (<23%).

These results suggest that significant proportions of small OMAs (<500 μm) form with ice blocks and slush ice in a relatively short period of time. As higher proportions of smaller aggregates formed with

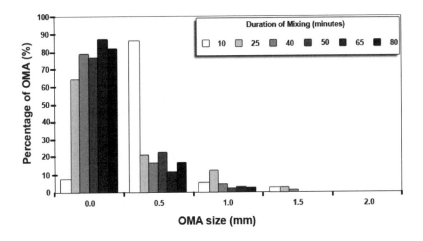

*Figure 13.* Distribution of the OMA size formed with ice blocks.

ice blocks are found, the results also suggest that turbulence damping is thus more important with slush ice than with ice rubble. On the other hand, the fact that the number of OMAs formed with slush is more elevated suggests that this particular type of ice may trap aggregates in the ice matrix. Laboratory experiments on OMA interaction with frazil ice were carried out by Payne *et al.* (1989). Their results suggest that oil contaminated sediment (mixed prior to their interaction with ice) may be incorporated into the ice cover during storm events (high energy level). These results clearly show the slush ice scavenging potential. Therefore, slush ice could be a potential source of OMA transport and dispersal.

3.6.  THE CASE OF OMA SIZE AFTER 12 DAYS

OMAs formed with slush ice were left in the basin for 12 days. It was observed that most of the aggregates remained in suspension after that period of time. An analysis of the size of particles left in suspend-ion was performed (Figures 14 and 15). Figure 14 shows the OMAs in suspension near the water surface. It is worth noting that OMAs of different sizes are still floating after that period of time. The analysis

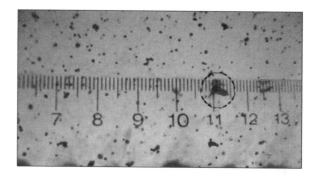

*Figure 14.* OMAs remaining in suspension in the basin 12 days after the tests. Mega-aggregates made of several distinct droplet OMAs are observed (red circle).

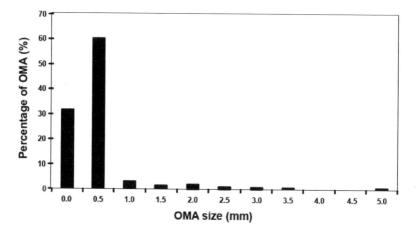

*Figure 15.* Distribution of the OMA size in suspension 12 days after formation.

of the particle size is presented at Figure 15. The results show a similar trend to that observed previously with OMA size distribution. In fact, the most important proportion of OMAs remaining in suspension is smaller that 500 μm (30% of OMAs < 0.5 and 60% of OMAs < 1.0 mm). t should be noted that OMAs as large as 5.0 mm are found in suspension after 12 days. These are multiple droplet OMAs formed of several distinct droplet OMAs. These observations demonstrate the large size range of OMAs formed with slush ice.

## 4.  Conslusions

Regarding the OMA process in ice-covered waters, the experiments showed that as soon as the oil is poured on ice, oil tends to migrate onto the ice matrix when slush ice is present and between ice rubble, forming oil droplets at the slick margin. When stirring energy is introduced, oil droplets form readily for all energy levels tested. Oil-mineral aggregates formed instantly with both ice types tested. Droplet OMAs were the only OMA type formed during the tests. The major proportion of OMAs formed were of small sizes (<1.0 mm) with both slush and ice blocks. The most efficient mixing period to form a significant amount of OMAs (~50%) is between 20 and 30 min, although a significant proportion of OMAs were observed after 10 min with slush ice. Slush ice seems to act as a filter that traps oil droplets as well as OMAs. OMAs formed in the presence of ice are stable and can remain in suspension for several days. These results suggest that the development of a procedure based on the formation of OMAs and aimed at dispersing oil spilled in ice-covered waters looks promising.

## References

Bragg, J.R. and Yang, S.H., 1995, Clay-Oil Flocculation and its Role in Natural Cleansing in Prince William Sound following the Exxon Valdez Oil Spill. *In Exxon Valdez Oil Spill: Fate and Effects in Alaskan Waters*, P.G. Wells, J.N. Butler and Hughes, J.S. Eds., American Society for Testing and Materials, Philadelphia, PA, pp. 178–214.

Cloutier, D., 2004. *Synthèse bibliographique et expertise technique sur le processus d'agrégation argile-pétrole dans les eaux infestées de glace*. Report submitted to the Hydraulic Engineering Sector, for the Environmental Intervention Division of Maritime Services, Quebec Region.

Cloutier, D., Amos, C.L., Hill, P.R. and Lee, K., 2003, Oil Erosion in an Annular Flume by Seawater of varying Turbidities: A Critical bed Shear Stress Approach. *Spill Science and Technology Bulletin*, 8(3): 83–93.

Delvigne, G.A.L., van der Stel, J.A. and Sweeney, C.E., 1987, *Measurement of Vertical Dispersion and Diffusion of Oil Droplets and Oiled Particles*. Delft Hydraulics Laboratory, Delft, The Netherlands, Report No. Z 75-2.

Guénette, C.C., Sergy, G.A., Owens, E.H., Prince, R.C. and Lee, K., 2003, Experimental Design of the Svalbard Shoreline Field Trials. *Spill Science and Technology Bulletin*, 8(3): 245–256.

Khelifa, A. and Lee, K., 2004, *Validation de la formation d'agrégats pétrole-argile dans une eau saumâtre et froide*. Research report submitted to the CCG, Hydraulic Engineering Sector, for the Environmental Intervention Division of Maritime Services, Quebec Region, 28 pp.

Khelifa, A., Stoffyn-Egli, P., Hill, P.S. and Lee, K., 2002, Characteristics of Oil Droplets Stabilized by Mineral Particles: Effects of Oil type and Temperature.*Spill Science and Technology Bulletin*, **8**(3): 19–30.

Lee, K., 2002, *Dispersion of Oil Spills Stranded in Ice and its Environmental Fate*. Report to Canadian Coast Guard.

Lee, K., Lunel, T., Wood, P. and Stoffyn-Egli, P., 1997, Shoreline Cleanup by Acceleration of Clay-Oil Flocculation Processes, in: *Proceedings of the 1997 International Oil Spill Conference*, American Petroleum Institute, Washington, DC, pp. 235–240.

Lee, K., Stoffyn-Egli, P., Tremblay, G.H., Owens, E.H., Sergy, G.A., Guénette, C.C. and Prince, R.C., 2003, Oil-Mineral Aggregate Formation on Oiled Beaches: Natural Attenuation and Sediment Relocation. *Spill Science and Technology Bulletin*, **8**(3): 285–296.

Payne, J.R., Clayton, J.R. Jr., McNabb, G.D. Jr., Kirstein, B.E., Clary, C.L., Redding, R.T., Evans, J.S., Reimnitz, E. and Kempema, E., 1989, *Oil-ice sediment interactions during freeze up and breakup*. Final Reports of Principal Investigators, U.S. Dept. Commer., NOAA, OCSEAP Final Rep, 64 pp.

Sergy, G.A. and Goodman, R., 2003, In situ Treatment and Fate of Oil Stranded on Coarse-Sediment Shorelines: the Svalbard Shoreline Field Trials. *Spill Science and Technology Bulletin*, **8**(3): 229.

Sergy, G.A., Guénette, C.C., Owens, E.H., Prince, R.C. and Lee, K., 2003, *In-situ* Treatment of Oiled Sediment Shorelines. *Spill Science and Technology Bulletin*, **8**(3): 237–244.

Weise, A.M., Nalewajko, C. and Lee, K., 1997, Oil-Mineral Fine Interactions Facilitate Oil Biodegradation in Seawater. *Environmental Technology*, **20**: 811–824.

# JOINT INDUSTRY PROGRAM ON OIL SPILL IN ARCTIC AND ICE INFESTED WATERS: AN OVERVIEW

## S.E. SØRSTRØM[†], I. SINGSAAS & P.J. BRANDVIK
*SINTEF Materials and Chemistry, Environmental Technology Department, NO-7465 Trondheim, Norway*

**Abstract[*].** SINTEF has on behalf of the oil companies Shell, Chevron, Statoil, Total and ConocoPhillips performed a pre-project for development of a R&D program for oil spill response in ice-infested and Arctic waters. AGIP KCO joined the program in 2006. The objective of the pre-project has been to propose objectives, scope of work and participants for a Joint Industry Program to develop tools and technologies for environmental beneficial oil spill response strategies in ice-infested waters. The main tasks in the pre-project have been:

- Preparation of a state-of-the-art report to give an overview of the R&D status within this field and form a basis for identification of future research needs as input to the JIP.
- Preparation of Joint Industry Program (JIP) proposal.

In addition to the funding oil companies provide, a number of co-operating organisations have agreed to join the program. The program will be presented during the third NATO/CCMS workshop on Oil Spill Response.

**Keywords:** oil spill response, arctic, ice, Joint Industry Program

---

[†] To whom correspondence should be addressed. E-mail: Stein.E.Sorstrom@sintef.no
. Full presentation available in PDF format on CD insert.

W. F. Davidson, K. Lee and A. Cogswell (eds.), *Oil Spill Response: A Global Perspective.*   89
© Springer Science + Business Media B.V. 2008

# COUNTERMEASURES FOR THE BEAUFORT TRANSITION SEASON

L. SOLSBERG[†]

*Counterspil Research Inc., 205–1075 West 1st Street,
North Vancouver, British Columbia, V7P 3T4, Canada*

**Abstract.** Spring breakup and fall freeze-up in the Beaufort Sea present unique oil spill response challenges that are often addressed in theoretical terms. The literature, in fact, has repeated over the years a basis for selecting countermeasures that might not reflect actual field conditions. Unfortunately, this approach will not likely assist those tasked with planning and implementing a response operation. More helpful decision factors are proposed that focus on the issues that require clarification and the options that need to be very quickly considered. To this end, insights are provided into oil and ice conditions, spill containment and oil removal. This information should not only allow more practical and effective decisions to be made by first responders but may also result in applied research in the future that answers some much needed questions.

**Keywords:** arctic, ice, oil spill response

## 1. Introduction

My work since 1973 has focused on spill response equipment and techniques, often related to oil in ice. More recent projects have included assessments in 2001 and 2006 of the mechanical response capability on Alaska's North Slope for BP Exploration (Alaska) Inc., instruction in cold weather spills with Polaris Applied Sciences, Inc. and Alaska Clean Seas to Chevron's World-wide Response Team (Solsberg and Owens, 2001), and with Polaris and Environment Canada, the development of a Field Guide for Oil Spill Response in Arctic Waters (1998).

---

[†] To whom correspondence should be addressed. E-mail: mail@counterspil.com

W. F. Davidson, K. Lee and A. Cogswell (eds.), *Oil Spill Response: A Global Perspective.*   91
© Springer Science + Business Media B.V. 2008

From 1995 to 2000, I also participated in SINTEF's Mechanical Oil Removal in Ice Program (MORICE). In 2006, BP also requested a study of *in-situ* burning for dealing with oil releases in the Beaufort Sea during the transition seasons. Numerous other projects have included contingency planning, training, and the review of response options for oil spills in ice in the Canadian and US North as well as the North Caspian Sea and Sakhalin Island.

Most of this work has indicated that for both oil spill response planning and safe operations to effectively proceed, oil and ice behaviour in relation to each other requires a clear understanding. It is the one complex data set that determines how, and really if, an oil spill can be controlled. At no time is oil/ice movement more dynamic than during fall freeze-up and spring breakup in the Beaufort Sea. In this paper, various factors are addressed relevant to applying mechanical equipment and burning to spills during the transition seasons. Advances in equipment are briefly summarized as well as the parameters that affect burning. Perhaps the first step in making responsible decisions has not yet been fully taken. Oil/ice studies are therefore proposed to resolve these shortcomings. The studies should result in knowledge of direct benefit to contingency planners and first responders.

## 2. Oil/Ice Behaviour

Traditional views of oil and ice have largely been based on two or three-dimensional depictions, i.e., they consider the x, y and often z axes as shown in Figure 1.

The primary pathways that spilled oil takes on, under and within ice are well known. Oil can initially concentrate between floes and tends to form pockets in, under, and also ultimately as melt pools on ice that might mean its removal is possible. Nonetheless, if the dimension of time is added, this picture becomes a dynamic one, especially during the transition seasons which generally exist in the nearshore waters of the Beaufort Sea during June-July and September-October. The implications to spills and countermeasures are significant because of the changes that can be expected to occur in oil and ice during such periods:

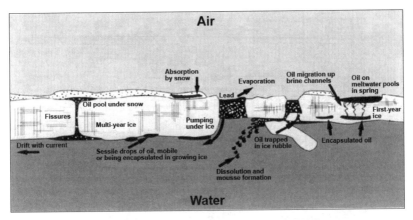

*Figure 1*. Oil behaviour in ice (Evers *et al.*, 2004).

- Ice size, shape, thickness, concentration and velocity are not static.
- Wind shifts in direction and speed are frequent.
- Wind largely determines ice position as well as oil trajectory.
- Oil spilled into ice will likely interact with it.
- The relative position of oil and ice will change over time.

The above factors require that first responders will have to make decisions much more quickly than they would for open water spills. This is because the conditions they plan for may change in a matter of hours (or less). Basic planning for response operations requires estimations of the extent of the ice cover and its characterization – the amount and type of ice present dictates what approach can be tried to control a spill and the safety precautions that must be taken. Mathematically modelling ice movement as it freezes or melts – and moves – is difficult and might not always be accurate. It will not likely contribute to improving either planning or engaging in an actual response.

When numerous floes are present with relatively few areas of open water that together can span distances of several kilometers, there is generally little doubt about ice cover – at between 90% and 100%, the ice cover can be readily estimated. Should a spill occur in such conditions, this still requires characterizing the oil in terms of its volume, thickness, viscosity, etc. Nonetheless, as the ice cover diminishes, assessments are not as easy for a number of reasons (see Figure 2):

*Figure 2*. Estimating ice cover is not an exact science.

- The observer's position determines oil/ice reporting numbers and can depend on height and distance; several different vantage points are necessary.

- Submerged ice and ice with sediment on it will reduce estimates of ice cover – not all of the ice that might affect the response will be visible.

- Oil might initially concentrate at an ice "edge".

- An ice "edge" might consist of a series of floes that will vary greatly in configuration and extend from only several metres to much longer distances.

- Oil will likely accumulate in small and large pockets.

- The ice can be expected to move with the spilled oil and form many small pools that change in character and move relative to each other as winds blow the oil/ice first in one direction and then another.

What we do know is that the spilled oil, if present in sufficient quantities, will likely initially collect in embayments and smaller pockets along the ice edges where it would be more amenable to removal. This is partially because oil and water appear to move more quickly than ice floes with oil slicks advected by winds of 5–10 km/h or more against the ice. Oil/ice tracking using spill simulators and ice tracking buoys will likely be of more use than mathematical modeling to determine trajectories – but applied studies are lacking. During a spill, airboats and aerial observation platforms, especially helicopters, will provide the key means to determine where the oil is relative to the ice, in what quantities, and if, and by what means, responders can remove it. In some cases, when oil has been released and has widely dispersed as fine droplets and/or the ice is present as very small pieces, or brash ice, no response measure may be practical. Whatever the circumstances, safety of operations will be, of course, the highest priority.

## 3.  Countermeasures for Oil in Ice

The response to oil spills in ice has been widely studied since the 1970s. A common way of presenting the potential use of countermeasures in Arctic conditions is a chart or table that assigns the standard approaches of mechanical operations, *in-situ* burning and dispersant application to a specific range of ice cover (Table 1). The problem with this approach, however, is that conditions in Beaufort Sea ice are so dynamic during the transition seasons that any response would likely prove to be limited to a small (time) window of opportunity – leave alone the obvious safety concerns. The same can be said for dealing with oil in moving ice in other parts of the world.

While specific ice cover limits for countermeasures are not exact, nor is precision always needed or applicable, what becomes apparent when in actual field conditions, is the utility of either platforms that can work at ice edges or a means to control spills using aerial methods. In either case, however accomplished, working outside of the collections of ice floes affords more opportunity for oil removal. To this end, burning oil appears to be the most practical response option with the widest range of possible application. Still, questions relating to any countermeasures for oil in ice point to the need to further understand ice drift and oil movement (so that we determine if oil is in fact going to be amenable to removal) as well as to the need to conduct applied studies of the various spill response technologies (to see what really works). There are existing sources of information that can help as regards the latter.

TABLE 1. Countermeasures for oil in ice.

| | Ice cover | | | | | | | | | | |
|---|---|---|---|---|---|---|---|---|---|---|---|
| | 0% | 10% | 20% | 30% | 40% | 50% | 60% | 70% | 80% | 90% | 100% |
| **Mechanical** | | | | | | | | | | | |
| **Booms & skimmers** | ▬▬▬ | ▬▬▬ | ▬▬▬ | ▬ | | | | | | | |
| **Vessel & skimmers** | | ▬▬▬ | ▬▬▬ | ▬▬▬ | ▬▬▬ | ▬▬▬ | ▬▬▬ | ▬▬▬ | ▬ | | |
| **Skimming vessels** | | ▬▬▬ | ▬▬▬ | ▬▬▬ | ▬▬▬ | ▬▬▬ | ▬▬▬ | ▬▬▬ | ▬▬▬ | | |
| **ISB** | | | | | | | | | | | |
| **Fire booms** | | ▬ | ▬▬▬ | ▬▬▬ | ▬ | | | | | | |
| **Burns in ice** | | ▬ | ▬▬▬ | ▬▬▬ | ▬▬▬ | ▬▬▬ | ▬▬▬ | ▬▬▬ | ▬▬▬ | ▬▬▬ | ▬ |
| **Dispersants** | | | | | | | | | | | |
| **FW Aircraft** | ▬▬▬ | ▬▬▬ | ▬▬▬ | ▬ | | | | | | | |
| **Helicopter** | | ▬▬▬ | ▬▬▬ | ▬▬▬ | ▬▬▬ | ▬ | | | | | |
| **Boat** | ▬▬▬ | ▬▬▬ | ▬ | | | | | | | | |

*Figure 3.* Field Guide for Oil Spill Response in Arctic Waters (Owens *et al.*, 1998).

### 3.1.  A FIELD GUIDE FOR OIL SPILL RESPONSE IN ARCTIC WATERS

A *Field Guide for Oil Spill Response in Arctic Waters* (Owens *et al.*, 1998) (Figure 3) was completed for the eight-nation Emergency Prevention, Preparedness and Response (EPPR) Working Group. The *Field Guide* is intended for use by the circumpolar countries to facilitate oil spill sresponse for seas, lakes, rivers, and shorelines. It addresses the unique climatic and physiographic features of the Arctic environment. Practical information is presented for technical managers and decision makers as well as local community first responders. The *Field Guide* does not duplicate existing manuals and documents, but rather, collates available information on the behaviour of, and response to, oil in ice and snow conditions.

"Part A" of the *Field Guide* is comprised of three key operational sections that can be used independently:

1. **A field guide for first responders** based on seasonal conditions that focuses on practical actions that can be taken to control the spread of oil and minimize its effects

2. **Response strategies** described in the context of source control, control of free oil, protection, and shoreline treatment and

3. **Response methods** for on water, in ice, and shoreline treatment.

"Part B" of the *Field Guide* contains technical support information basic to understanding the Arctic environment and to developing an effective response. It addresses the **behaviour and fate** of spilled oil and the **notification and decision processes** associated with managing a response operation. A final section summarizes the **coastal character** of each Arctic geographic region.

The *Field Guide* presents practical information in an original style and format. For example, bullets and icons are used throughout the *Field Guide* because they can be recognized internationally, regardless of language. General approaches are presented rather than the technical specifications or details of equipment and methods, e.g., those being researched via MORICE.

While very comprehensive, the limitations of the *Field Guide* are similar to those of other references. Response options for the transition seasons are generally well indicated for oil in ice but the details of oil movement in relation to ice drift and the need for quick decisions are not fully discussed. It is these aspects, as well as technology updates, that would add to the extent and practicality of the information that has been compiled. The versatility of burning oil in ice is very apparent when consulting countermeasures options in the *Field Guide.*

### 3.2. BEST AVAILABLE TECHNOLOGY (BAT) ANALYSES

Best Available Technology (BAT) analyses were conducted in 2001 and 2006 on behalf of BP Exploration (Alaska) Inc. for mechanical oil spill response equipment in North Slope marine ice during the transition seasons following regulations developed by the Alaska Department of Environmental Conservation (ADEC). The eight BAT criteria are as follows (Solsberg *et al.*, 2002):

1. **Availability**

   Is the technology alternative being considered the best in use in other similar situations and is it available for use by the applicant? Consider technologies used in other industries world-wide.

2. **Transferability**

   Is the technology alternative being considered transferable to the applicant's operations? Is it applicable to this situation?

3. **Effectiveness**

   Is there reasonable expectation that each technology will provide increased spill prevention or other environmental benefits? In this context, will the technology be more effective in containing a discharge and decreasing impacts to air and water.

4. **Cost**

   What is the cost to the applicant of best available technology, including its cost relative to remaining years of service of currently used equipment?

5. **Age and condition**

   What is the age/condition of existing versus available technologies.

6. **Compatibility**

   Is each technology compatible with existing operations and technologies?

7. **Feasibility**

   Is it feasible to use a particular device/system, technically and operationally?

8. **Environmental impacts**

   Will impacts to air, land, and water offset any environmental benefits?

Three analyses of response equipment were conducted for each freeze-up and breakup in which the eight BAT criteria were applied to (1) commercially available spill technologies, (2) a limited number of devices not yet generally available, and (3) additional promising technologies. MORICE equipment was reviewed (from SINTEF's Mechanical Recovery of Oil in Ice program), as well as other concepts (primarily from Lamor Corporation Ab of Finland) specifically developed for the recovery of oil in ice.

Equipment studied included at least three each of small and large booms, skimmers and pumps as well as various vessels and barges. Vessel-skimmer-boom containment-and-recovery systems were also examined for use in ice near the Realistic Maximum Response Operating Limitations (RMROLs). U-boom and various skimmer systems were also evaluated for their operation in narrow leads and in ice fields. Skimming systems independent of booms were also studied.

Where equipment capability varies as a function of oil properties, then this distinction was made. For example, oil thickness and footprint differs between batch spills and blowout oil deposited as droplets. In addition, the degree of oil weathering, icing problems, and the operation of ancillaries (e.g., scrapers, combs, wringers) as these might affect performance were factored into the evaluations.

It is important to note that the BAT reviews utilize a methodology that could be applied by other organizations to evaluate and document realistic oil-in-ice response capabilities, including equipment and strategies, for operators, regulators and stakeholders. BAT also provides a sound basis to determine R&D needs since limitations become evident from a number of different perspectives.

### 3.3. RESULTS OF RESPONSE EQUIPMENT REVIEWS

Large, heavy-duty booms are commercially available that have a proven record in harsh offshore environments, excellent stability, and design features suitable for potential use in ice infestations. Their limitations in ice include disruption of the oil-ice mixture, possible damage, and inaccessibility of skimmers to the oil or their inability to process the oil/ice mixture presented by the boom.

Small containment booms are more suited to open water response operations than to operations in ice. BAT, in terms of a small booms could consider however, boom dimensions, and fabric details (e.g., tensile and tear strength, cold crack temperature, flexibility, crush-resistant flotation, fittings, etc.)

Many large high volume pumps are available that are suited for transferring fresh and emulsified crude oil from tanks but were not designed to move slush ice and oil. Enhancement of screw auger pumps (Figure 4) using steam/heat and annular flow is now considered to be available technology (Cooper, 2004). Advances made over the past several years have been significant. While such pumps incorporate components that have been improved for processing debris and viscous

*Figure 4.* Screw auger pump.

*Figure 5.* Brush drum on arm.

oil (e.g., multiple cutting knives, split casing, redundant sealing discs, and Teflon-impregnated metal) they still can encounter problems transferring slush and small ice pieces at sub-freezing temperatures.

Various small pumps were considered to be BAT that have potential application to skimming and general transfer duties for water, oil and small ice forms. Eliminating elbows, indirect feed, limiting exposure to sub-freezing temperatures before use, draining between uses, complete contact of electrical connections, etc., contribute to good pumping capability. Still, there are many problems present when small pumps are used in winter conditions.

For operation during freeze-up and breakup on the North Slope, the barges Arctic Endeavor and Beaufort 20 are well suited for their assigned tasks. Similarly, the ACS Bay Boats are considered to be Best Available Technology suitable for available response equipment and its capabilities.

In a continuous ice cover, the brush drum, brush adaptors and brush pack skimmers comprise the best choices for dealing with oil spills, but still have limited capability. Not all oil present will likely be removed. The brush drum skimmers attached to a hydraulic arm (Figure 5) can be readily positioned in oil in ice, operated there, and then lifted out and moved elsewhere as needed. These skimmers should be supplemented with other options including vertical mop and weir skimmers that could be used when *stationary* oil collection is possible in ice that does not significantly affect their performance.

Their advantage is that ice pieces are processed under the recovery device so that ice does not clog the system. However, there is a loss of some of the oil as it passes under the device along with the ice. During freeze-up, it is expected that the drum brush might marginally increase the ice concentration that an oil recovery system could function in, but would miss oil as it processes ice. Lamor Corporation Ab manufactures brush belts and adapters (Figures 6 and 7).

Even with the advances that have been made recently with skimmers and their ability to process small ice pieces, the challenges that remain for mechanically removing oil from moving ice can be summarized as follows:

*Figure 6.* Brush belt.

*Figure 7.* Brush adaptor.

- Limited access to oil
- Reduced oil flow to the skimmer
- Icing/freezing/jamming of equipment
- Separation of oil from ice
- Contamination/cleaning of ice
- Deflection of oil together with ice
- Strength and durability considerations
- Detection, monitoring of slick.

Applying multiple smaller skimmers in ice might be possible if air temperature is above freezing and if there is access to discrete pockets of oil. Slick thickness should be about 2.5 cm (1 in.) for significant recovery for blowouts or batch releases that have been previously contained by booms and/or ice.

The other primary approach to skimming oil in ice has evolved into advancing concepts that either raise ice pieces (MORICE) or that deflect them downward (Jensen *et al.*, 2002; Jensen and Solsberg, 2001). Oil is then separated and collected. In the case of MORICE, the oil is washed off ice that passes over a grated belt and is picked up by rotating brush drums while Lamor's Oil Ice Separator (LOIS) features a vibrating grate that deflects ice downward and then recovers oil via a brush pack (Figures 8, 9 and 10).

*Figure 8.* Testing in oil/ice.

*Figure 9.* MORICE.

*Figure 10.* Lamor LOIS.

Both the MORICE and LOIS approaches are predicated on the following scenario:

- Ice pieces 0 to 2–3 m ice size (*not* large floes)
- Sixty to 70% ice concentration on large scale, locally up to 100%
- Brash and slush ice present
- Moderate dynamic conditions (waves, wind, current).

There might be specific situations where mechanical systems should be further investigated; however, many systems have potential limitations related to ice processing problems, air temperature (freezing), or overall practicality and effectiveness in moving transition season ice:

- Various rope mop configurations
- Air and water jets to deflect oil under ice
- Improvements to Transrec (including an Arctic model)
- Ice boom to manage ice – and perhaps oil
- Continued work on augers (as a component of other systems)
- Various processing equipment (from other applications, e.g., mining).

*Figure 11.* LAS skimmer.

Lamor Corporation (2006) has developed several iterations of a device that now incorporates components that have much greater potential for collecting oil, albeit in a stationary mode. Lamor's Arctic Skimmer (LAS) (Figure 11) utilizes two brush wheels that rotate in the same direction as water flow to force oil under water. The water flows out holes at the back of the skimmer while the oil adheres to the brushes and enters a hopper. Screws in the hopper crush ice pieces and feed oil, ice and slush to a built-in screw auger pump.

## 3.4. *IN-SITU* BURNING

Because oil can be removed using a Helitorch (Figure 12), *in-situ* burning comprises a response option that can be quickly and safely implemented for a variety of ice conditions (Fingas and Punt, 2000). The oil must still be present in a burnable state and quantities but a much wider range of ice cover can be addressed that will not interfere with operations as they would with mechanical methods.

*Figure 12.* Helitorch.

While fire booms might not always have application as ice cover increases, similar to their conventional open water counterparts, the concentrating and holding effects of ice floes could make burning feasible. Again, the relative behaviour of spilled oil and ice require further investigation to determine how feasible it is, in fact, to burn oil during the transition seasons in dynamic ice. Demuslifiers and herders have been studied with some degree of potential success in this regard. However, this work has been conducted in smaller scale tests where edge effects and other factors might result in burns that would not otherwise occur. While small and meso-scale testing is needed, field work is the ultimate means of measuring the feasibility of burning.

The key questions that arise when considering burning oil in an ice environment relate to the following issues:

- Dealing with the quickly changing oil and ice conditions of the Beaufort Sea during breakup and freeze-up.

- Determining the location, amount, burnability of oil (API gravity, oil thickness, % water in the oil – emulsification, weight % distribution).

- Ability to quickly deploy personnel and resources on sufficient scale so as to result in significant oil removal rates.
- The most effective means of utilizing the Helitorch and perhaps modifying it and locating burnable quantities of oil during a sortie.
- Determining the nature of the residue as to whether it floats or sinks and its potential impacts? (Net Environmental Benefit Analysis).
- Investigating the effects of ice type on burn, especially in terms of smaller formats such as frazil, brash, etc.
- Assessing in field conditions the practicality of applying firebooms, demulsifiers and herders.
- Safety of operations.

## 4.  Conclusions

The conclusions and recommendations that can be derived from the various projects conducted for oil-in-ice since 1990, particularly for the transition seasons in the Beaufort Sea, can be summarized as the following R&D and response priorities:

- Study the tracking of oil movement relative to ice drift by engaging in field work using spill simulators and tracking buoys if actual oil cannot be used.
- Select response tactics that may be effective for short periods based on field observations that utilize various vantage points.
- Plan for the quick implementation of countermeasures at strategic locations that might only be effective for several hours at a time.
- Focus on responder needs that accurate field information yields and not generalized applications indicated by charts and test tank data found in the literature.
- *In-situ* burning is likely the key oil removal method and requires further study of aerial application methods and oil/ice movement so that even higher efficiencies might result.
- Know details of oil and ice interactions and effects on spills for specific geographic locations when planning response operations.
- SAFETY of operations is the Number One Priority.

## 5.  Acknowledgements

Alaska Clean Seas, SINTEF, Lamor Corporation Ab and BP Exploration (Alaska) Inc. are acknowledged for the insights that project work and discussions with company personnel have provided. Various operational issues related to using equipment in cold weather and modifications that the organizations had considered were discussed. Equipment manufacturers supplied technical specifications and test data as requested. In the 2001 work for BP, ADEC clarified BAT requirements and requested additional information that resulted in further insights into steam enhancement, the use of pumps and large booms, and the potential application of specific skimmer systems.

## References

Cooper, D., 2004, Testing of a Modified GT260 Pump in Viscous Oil, SAIC Report.

Evers, K.-U., Jensen, H., Resby, J., Ramstad, S., Singsaas, I., Dieckmann, G. and Gerdes, B., July 2004, State of the Art Report on Oil Weathering and of the Effectiveness of Response Alternatives, ARCOP Report D4.2.1.1(a).

Fingas, M. and Punt, M., February 2000, *In-situ* Burning, A Cleanup Technique for Oil Spills on Water.

Jensen, H.V. and Solsberg, L., 2001, The Program for Mechanical Oil Recovery in Ice-infested Waters (MORICE), Phase 5, SINTEF Report unassigned, Lamor Corporation Ab Literature, 2006, Lamor Brush Arctic Skimmer LAS 125 W/P, LOIS and LRB Skimmer.

Jensen, H.V., Mullin, J.V. and McHale, J., 2002, MORICE Oil in Ice Testing at Ohmsett, Proceedings from the Twenty-fifth AMOP Technical Seminar, pp. 1269–1283.

National Oceanic and Atmospheric Administration (NOAA) website.

Owens, E.H., Solsberg, L.B., West, M.R. and McGrath, M., Field Guide for Oil Spill Response in Arctic Waters, September 1998, Emergency Prevention, Preparedness and Response Working Group, published by Environment Canada, 348 pp.

Solsberg, L. and Owens, E.H., June 2002, Cold Weather Oil Response Training, Proceedings of the 24th Arctic and Marine Oilspill Program AMOP, Technical Seminar, Edmonton, Canada, pp. 187–194.

Solsberg, L., Glover, N.W. and Bronson, M.T., June 2002, Best Available Technology for Oil in Ice, in: Proceedings of the 25th Arctic and Marine Oilspill Program AMOP, Technical Seminar, Calgary, Canada, pp. 1207–1224.

# PART 2. OIL SPILL COUNTERMEASURES

# NEW OIL CONTAINMENT TECHNOLOGY: FOR FAST AND NARROW WATERS

J. ALLERS
*AllMaritim AS, Box 51, 5812 Bergen, Norway*

D. TRITES[†]
*DSS Marine Inc., 71 Wright Ave., Dartmouth, Nova Scotia, Canada*

**Abstract**[*]. It is a well known fact that conventional oil booms towed in U-, J- as well as V-configurations will lose oil when towed in more than 0.9 knot through the water. The low towing speed makes boom towing a challenging exercise for any kind of dedicated or non-dedicated pair of towing vessels. Towing in narrow waters (or in between ice patches) becomes a futile effort when one has to make sharp turns with oil in the boom. A turn usually causes an acceleration of the outside boom arm – and the oil contained gets lost! This paper will discuss how new oil containment technology, commercially known as the Buster Technology, has made both boom towing in fast waters and quick manoeuvring fully acceptable. Following extensive testing in oil in the world's largest onshore test tank, the Current Buster has since been involved in four real life oil spills ranging from diesel to heavy fuel oil. One of the spills took place in very narrow waters and clearly demonstrated the usefulness of the system's high manoeuvrability. And in two of the spills oil was safely contained and controlled at towing speeds of 3.5 knots.

**Keywords:** oil spill response, current buster

[†] To whom correspondence should be addressed.  E-mail: dtrites@dssmarine.com
[*] Full presentation available in PDF format on CD insert.

W. F. Davidson, K. Lee and A. Cogswell (eds.), *Oil Spill Response: A Global Perspective.*    113
© Springer Science + Business Media B.V. 2008

# VISCOUS OIL PUMPING TECHNOLOGY AND ANNULAR WATER LUBRICATION TECHNIQUES

J.P. MACKEY[†]
*Hyde Marine Inc, 28045 Ranney Parkway, Cleveland OH 44145, USA*

**Abstract[*].** Great advancements have been made with the tools and techniques available for responders to pump extremely high viscosity oils and emulsions. The Joint Viscous Oil Pumping System (JVOPS) workshop in December 2003 demonstrated to the oil spill response industry that existing pump technology could be used to transfer very high viscosity oils over operational distances and at safer discharge pressures by employing simple techniques and a relatively small investment in new technology. Industry has started to implement many of the recommended upgrades and improvements to the existing inventory. Modern pumps, with higher torque motors and integrated Annular Water Injection technology are making their way into the market. However, there is still work to be done to enable responders to get the maximum benefit from these techniques. This presentation will review the results from JVOPS and the current state of the art of positive displacement Archimedes' screw pump design and annular water lubrication systems. It will also highlight some areas in need of further development.

**Keywords:** pumping technology, oil spill response, lubrication techniques

---

[†] To whom correspondence should be addressed. E-mail: jim.mackey@lamor.com
[*] Full presentation available in PDF format on CD insert.

W. F. Davidson, K. Lee and A. Cogswell (eds.), *Oil Spill Response: A Global Perspective.*   115
© Springer Science + Business Media B.V. 2008

# DEVELOPMENT OF A STRATEGY FOR OFFSHORE USE OF DISPERSANTS IN NORWEGIAN WATERS

## I. SINGSAAS[†], M. REED & T. NORDTUG

*SINTEF Materials and Chemistry, Environmental Technology Department, NO-7465 Trondheim, Norway*

**Abstract**[*]. Oil spill contingency planning consists of evaluating potential discharge scenarios for the location in question and developing response strategies. The main objective for a response strategy is to minimize the environmental consequences of an oil spill on ecological, comer-cial and/or human used resources. The weighting of relative advantages and disadvantages and the study of consequences with use of different oil spill countermeasures is often referred to as Net Environmental Benefit Analysis. Modelling tools have been developed to give support for such analyses. SINTEF has performed oil spill contingency and NEBA analyses for the oil industry over a period of 10–15 years. The OSCAR (Oil Spill Contingency And Response) model was developed in the early 1990s to support such analyses and has been continuously strength-ened since then by, e.g., improving the simulations of water soluble oil components and dispersed oil droplets in the water column. Recently the model has been further developed from a scenario-based model to also allowing for stochastic simulations. OSCAR is a multi-component three-dimensional modelling tool used for analysing alternate response strategies. Key components in the system are: a data-based oil weathe-ring model; a near zone model; a three-dimensional oil drift model; a strategic response model; Exposure models for fish and plan-ktonic organisms, birds and marine mammals; and, tools for evaluation of exposure within GIS polygons. The model analyses alternate response strategies (e.g., mechanical recovery vs. use of dispersants) as a basis for a quantitative evaluation of environmental risk in the marine envi-ronment. The model has been used as a basis for evaluation and deve-lopment of strategies for use of dispersants in Norwegian waters, both offshore and for oil terminals and refineries. Dispersants can be used as a supplement to mechanical recovery or as an alternative in certain

---

[†] To whom correspondence should be addressed. E-mail: ivar.singsaas@sintef.no

[*] Full presentation available in PDF format on CD insert.

scenarios. The decision model for use of dispersants is based on the following criteria: Is natural dispersion already a dominating process?; Which biological resources are threatened by the oil spill and how will use of dispersants influence upon these?; Will the dispersed oil be effectively diluted in the water column?; Will the effectiveness be reduced due to oil type and/or weathering degree?; Will the effecttiveness be reduced due to low salinity (brackish water)?; Will the effectiveness be reduced due to bad weather (wind/fog)?; How to apply the dispersant in a correct manner?; Is there sufficiently short response time and treatment capacity?; How to monitor the effectiveness of the dispersant action?; and, Criteria for when and how to terminate the dispersant application. A strategy for use of dispersants has been developed for several Norwegian offshore oil fields based on this methodology. Restricted by the amount of dispersant available, dispersants can be used as an alternative to mechanical recovery for smaller oil spills (typically less than $500 - 1000$ m$^3$) contributing a supplement for larger oil spills.

**Keywords:** oil spill response, dispersants, SINTEF, Norway, model

# FRENCH SEA TRIALS ON CHEMICAL DISPERSION: DEPOL 04 & 05

F.-X. MERLIN[†]
*CEDRE, 715 r. A. Colas, CS: 41836, Brest Cedex 2,
F-29218, France*

**Abstract[*].** In 2004 and 2005, Cedre organized sea trials off the coast of Brittany with the French Navy and in collaboration with the French Customs. While the 2004 trials were large experiments looking for global assessment of the technique of dispersion, the 2005 sea trials were small scale sea trials focused on the efficiency of the dispersant product itself. The 2004 sea trials, DEPOL 04, involved three controlled oil discharges which were treated with two chemical dispersants using aerial spraying equipment, (Cessna equipped with a spraying POD) and shipborne spraying equipment. The slicks' evolution was monitored with remote sensing techniques, sampled for analysis and measured *in-situ* with spectrofluorometry. The objectives of these sea trials were:

- To study the natural weathering of the slicks
- To assess the chemical dispersion of the slicks
- To assess the operational possibilities of the spraying systems
- To run the annual *Bonnex* intercalibration exercise of the remote sensing means of the Bonn Agreement members
- To test new remote sensing devices under development
- To test roughly an oil recovery device purchased recently by the French Navy, (Sweeping Arm), to equip its spill control vessels.

The 2005 sea trials, DEPOL 05, aimed to establish an at sea testing procedure on small oil slicks to assess the real efficiency of dispersants versus different oil types. This paper presents these sea trials and their results considering mainly the chemical dispersion. For DEPOL 04, the dispersant treatment gave positive results despite the very calm meteorological conditions:

---

[†] To whom correspondence should be addressed. E-mail: francois.merlin@cedre.fr
[*] Full presentation available in PDF format on CD insert.

W. F. Davidson, K. Lee and A. Cogswell (eds.), *Oil Spill Response: A Global Perspective.*  119
© Springer Science + Business Media B.V. 2008

- Although the first slick was not totally treated with dispersant, most of the oil was dispersed.

- The comparison of the last two slicks' evolution tends to show either a significant advantage of the aerial treatment over the ship-borne one, or a higher efficiency of one dispersant over the other one.

For DEPOL 05, an operational incident forced the planned testing program to stop prematurely, but the proposed procedure proved to be promising: such a testing method will allow the running of an important number of comparative tests while working with controlled application conditions especially the dispersant-oil-ratio.

**Keywords:** dispersants, oil spill response, spraying, remote sensing

## 1. Introduction

In 2004 and 2005, CEDRE organized sea trials off the coast of Brittany with the French Navy and in collaboration with the French Customs.

While the 2004 trials were large experiments looking for global assessment of the technique of dispersion, the 2005 sea trials were small scale sea trials focused on the efficiency of the dispersant products.

The 2004 sea trials, DEPOL 04, involved three controlled oil discharges which were treated with two chemical dispersants using aerial spraying equipment, (Cessna equipped with a spaying POD) and ship borne spraying equipment. The slicks' evolution was monitored with remote sensing techniques, sampled for analysis and measured *in situ* with spectrofluorometry.

DEPOL 04 sea trials covered the following main objectives:

- Assessment of the chemical dispersion of the slicks, when treated with an aerial application system, the Cessna-POD, and with a ship-borne application system.

- Study of the weathering of paraffinic and asphaltenic oils.

These sea trials gave the opportunity to carry out additional tasks:

- The BONNEX exercise, intercalibration of the aerial remote sensing equipment of the Bonn Agreement members, (Sweden, Belgium, United Kingdom and France participated in this exercise).

- The testing of new oil detection devices under development, DETECSUIV (ACTIMAR French company), LIDAR (NMRI, Japan).

- Finally, the French Navy tested at sea its new recovery device, the sweeping arm which equips its supply vessels.

DEPOL 05 sea trials, aimed at setting a procedure to conduct comparative and well controlled tests in open sea, on small oil slicks, on different oils and dispersants, in order to assess the real efficiency of dispersants versus different oil types. The testing procedure was designed to obtain a good control of the oil and dispersant application conditions, especially the dispersant-oil-ratio.

These experiments have been carried out with the additional co-operation of TOTAL S.A. which supplied the oil and the dispersant, OSRL which owns the Cessna POD dispersant application system, and SINTEF (Norway) and MUMM (Belgium) for scientific support.

## 2.  DEPOL 04 General Organization

### 2.1.  GENERAL PROGRAMME

DEPOL 04 sea trials lasted over three days:

- First day: release of 10 m$^3$ of a paraffinic oil, slick A, which was left to weather for ~6 h and was treated with aerial application of dispersant.

- Second day: release of two 10 m$^3$ slicks of asphaltenic oil, slicks B and C, which were left to weather for 7 h; the slicks were then treated with dispersant either by aerial application or by shipborne application.

- The third day was devoted to the recovery of the residual oil.

### 2.2.  ANALYSES, MEASUREMENTS AND DATA COLLECTION

Different data collections were performed during these sea trials:

1.  Oil sampling of the slicks was carried out with rubber boats, in order to measure, in the onboard laboratory, the physical properties

of the oil (viscosity, density, emulsification) (Guyomarch *et al.*, 2001, 2002).

2.  Spectrofluorometry measurements were conducted with rubber boats in order to assess the dispersed oil content in the water column.

3.  Aerial imagery was carried out by the six remote sensing aircrafts, and two additional ones, flying over the slicks: visible, IR and UV, and laser fluorometry sensors were used and sometimes combined together.

As a general comment, it was quite difficult to carry out oil samp- ling on the slicks; the long distance between the slicks and the support vessel, as well as the operational restrictions due to the take off and landing of the helicopter on the support vessel, led to limitations for bringing back on board as many samples as expected.

In addition to that, the spectrofluorometry measurements did not give all expected information on the dispersed oil concentration in the water column; the operators faced different problems such as pollution of the measuring cell of the equipment which led to false measure- ments, bogging down of the internal memory of the equipment which did not allow downloading the recorded data.

The collection of oil samples as well as the measurements at sea met with difficulties due to logistical and technical problems. Thank- fully, the 8 aircraft which flew over the slicks collected a large amount of images which contributed to interpreting the behaviour of the slicks and the dispersion process.

## 3. Description of the Tests

### 3.1. METEOROLOGICAL SEA CONDITIONS

During the three days, the meteorological conditions were quite calm especially on the second day (see Table 1).

### 3.2. "SLICK A"

#### 3.2.1. *The Oil*

The oil was a mixture of fresh crude oils, especially North sea crude oils, chosen in order to get a significant proportion of paraffinic com- pounds. It was pre-weathered in CEDRE by evaporating 11% of its volume.

TABLE 1. Meteorological conditions during sea trials DEPOL 04.

|  | Wind (m/s) | Average (m/s) | Sea temperature |
|---|---|---|---|
| 05–25th- morning | 0 to 1 | 0.5 | |
| 05–25th- afternoon | 1 to 3 | 1.7 | |
| 05–25th to 26th- night | 3 to 2 | 2.6 | 15°C |
| 05–26th- morning | 2 to 1 | 1.3 | |
| 05–26th- afternoon | 1 to 0 | 0.3 | |
| 05–26th to 27th- night | 0 to 2 | 0.5 | |
| 05–27th- morning | 0 | 0 | |

The properties and composition of this pre-weathered oil are given in Table 2; the saturate fraction of the oil is the majority. A representation of the GC pattern of the initial oil is given in Figure 1 and shows the evaporation process which affected the linear alkanes up to C16.

TABLE 2. Properties of the oil in slick A.

| Density | | 0.843 @ 14°C |
|---|---|---|
| Viscosity | | 7 mPa.s @14°C |
| Composition | Saturates | 68.7% |
| | Aromatics | 25.8% |
| | Resins | 4.6% |
| | Asphaltenes | 0.9% |

### 3.2.2. Description of the Operation

In the morning, the 10 m$^3$ of oil were released crosswind for around 500 m. The dispersant application was performed around 6 h after the oil release, by the airplane Cessna of OSRL, equipped with a POD dispersant application device and guided by the UK spotting aircraft.

According to the airplane crew, five spraying runs were performed, and 1 m$^3$ of Finasol OSR62 dispersant was applied on the slick. By the end of the day, (8 h after the oil release) a complementary treatment was undertaken by the French Navy vessel Lynx with Gamlen OD 4000 to get rid of the residual surface oil. As some remaining oil was observed on sea surface two days after the oil was released, the residual patches were recovered by the oil spill recovery vessel Alcyon equipped with a sweeping arm and 1 m$^3$ of emulsion was collected.

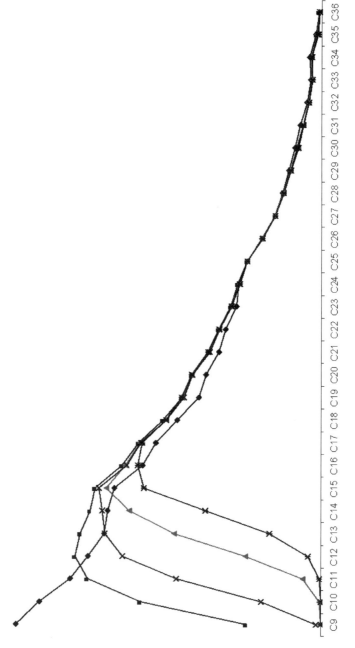

*Figure 1.* n-Alkanes distribution of slick A oil, at the time of the release and 1, 4, 7 and 45 hours after (all the GC are normalised on C25).

Figure 2 gives IR thermographies from the French Customs describing the evolution of slick A, after release, after dispersant application and one day later, respectively.

### 3.2.3. *Oil Behaviour*

One hour after the release, the oil viscosity reached 3,400 mPa.s and the water content of the oil was found to be around 80%; the emulsion stability increased progressively: at the beginning, 75% of the water settled after 2 h, just before treatment 20% settled after 2 h. Small lumps of emulsion (see Figure 3) gathered to form the thick part of the slick were reported by observers, which can be interpreted as the crystallisation of the paraffinic compounds.

At the time of the treatment the area of the slick was about 100–120 ha (3 × 0.4 Km). After treatment the viscosity dropped to 1,000 mPa.s, but the emulsion water content remained relatively high, around 66%; however, its stability decreased significantly: 40% water settled after 2 h. Observers who sampled the slick reported a clear reduction in the number of emulsified lumps; the residual patches tended to break up when subjected to some agitation (e.g., a ship bow wave). Two days post oil release, (T = 47 h), the remaining emulsion pre-sented water content of 59% (Figures 4 and 5).

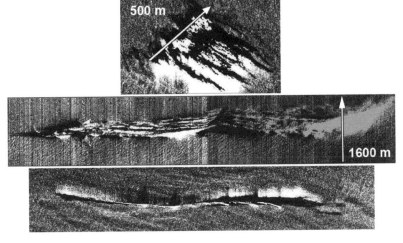

*Figure 2.* IR thermographies of slick A after release, after dispersant treatment and one day later.

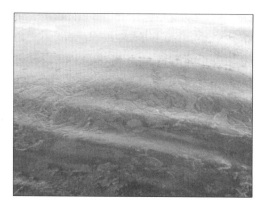

*Figure 3.* Appearance of the emulsion before treatment.

*Figures 4 and 5. (4)* Recovery of the remaining emulsion from slick A with the sweeping arm in very calm seas two days after release. *(5)* Aspect of the emulsion.

### 3.2.4. *Dispersant Application*

On aerial images the tracks of only four runs are clearly visible (1,900–3,040 m long and ~40 m width); the treated surface can be assessed to 38 ha, which represents between 30% and 40% of the total surface of the slick (see Figure 7).

The treatment led to the clear reduction of the area of the thickest part of the slick (which dropped from 6% to 3% of the total surface of the slick). Figure 6 shows a stripe of remaining emulsion after the treatment. In the area of the slick, the relative oil concentrations in the water column measured with Spectrofluorometry doubled after the dispersant application.

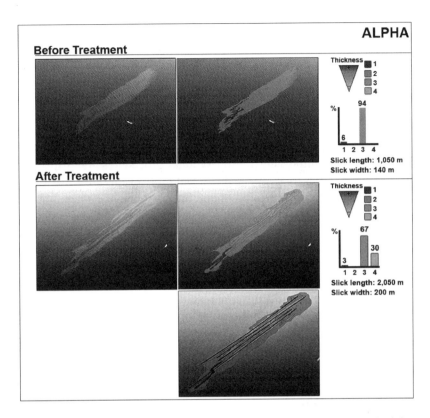

*Figure 6.* Visible picture of slick A before and after dispersant treatment: on the left – aerial photography; on the right – the original pictures have been processed to assess the relative areas covered by the different oil thickness (black emulsion to sheen) – see the histogram; the four treatment runs are indicated in the bottom picture.

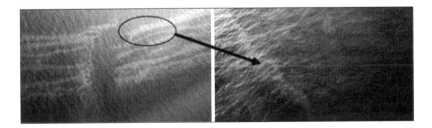

*Figure 7.* Appearance of the slick after treatment with some untreated stripes of emulsified oil.

### 3.2.5. *Discussion*

Despite the fact that the dispersant treatment of this slick was carried out under optimal conditions (good visibility, slick well targeted, assistance of the UK spotting aircraft for guiding, well trained crews), the slick was not totally treated (roughly half of the slick remained untreated as well as half of the thickest part – see Figure 7). The remaining emulsion recovered two days later could probably be attributed to the emulsion left untreated. These observations confirm that there are still possibilities for improving the operational procedures for dispersant application in order to apply the dispersant more evenly over the whole slick; thus, avoiding untreated areas. Surprisingly, the width of the treated tracks (30–40 m) is much larger than what could be expected from the spraying equipment Cessna-POD (8 m according to the crew, possibly up to 10–15 m), which proves that the dispersant had some herding effect on the oil slick.

Despite the lack of natural agitation, the chemical dispersion with the Finasol OSR62 gave a positive result: it succeeded in reducing the amount of surface oil; the cubic metre of residual emulsion recovered after two days (1 m$^3$ which represented around 0.6 m$^3$ of pure oil) should be compared to the 10 m$^3$ of oil initially released which would have become about 30 m$^3$ of emulsion.

### 3.3. SLICKS B AND C

### 3.3.1. *The Oil*

The oil used for slicks B and C was asphaltenic; it was a mixture of Heavy Fuel Oil (60%) and Light Cycle Oil (40%). These two products were mixed in CEDRE facilities to get a homogeneous mixture with the following properties: density of 0.948 and 4.9% of asphaltenes. Preparatory work carried out in laboratory showed that this oil could give, at sea, an emulsion up to 80% water with 5,900 cSt viscosity; such characteristics would have been suitable to test the operational efficiency of dispersion.

### 3.3.2. *Description of the Operations*

For each slick, 9 m$^3$ of oil were released early in the morning, cross-wind, over approximately 600 m in length. Slick B was marked out with a dye spot of rhodamine (red) and slick C released half an hour later was marked out with fluoresceine (green). The surface oil was sampled between 1 and 3 h after the release, then again from 5 to 7 h after release. The dispersant was then applied by the Cessna-POD on slick B and by the ship Lynx on slick C. After the dispersant application, ships were sent cruising at high speed on each slick to bring some mixing energy in order to enhance the dispersion process (Alcyon on slick B and Lynx on slick C). After an additional sampling session (T = 8.5 h), and an aerial evaluation (T = 10 h), a complementary dispersant application was undertaken to treat the residual oil. As previously, this complementary treatment was carried out by the Cessna POD on slick B and the ship Lynx on slick C; the dispersion application was followed with mixing provided by ships cruising in the slick.

### 3.3.3.  *Oil Behaviour*

The weather was very calm and the sea quite flat (wind speed ~1 m/s dropping to 0 during the afternoon); therefore, and unexpectedly, no real emulsification (formation of water in oil emulsion) was observed; sample water content was between 1% and 5% for slick B and in the range of 0–3% for slick C (Figure 8). On both slicks the surface oil sampling became almost impossible due to the rapid spreading of the oil on a large area combined with the lack of emulsification which resulted in very low thicknesses; therefore, for the following sampling sessions it was too difficult for the dinghies to collect enough oil to get a significant sample.

### 3.3.4.  *Dispersant Application*

The objective of this trial was to compare the different application methods, (airborne and shipborne application) with mixing energy brought by ships cruising in the slicks to take into account the calm calm sea conditions; however, while the Cessna POD used Finasol OSR 62 on slick B, the ship on slick C had to apply another dispersant, Gamlen OD 4000, due to an unexpected technical glitch which would not allow a proper connection to spraying equipment on the special additional tank rigged on the deck containing the FINALSOL OSR62. Despite these difficulties, observations from aerial images of the slicks gave some interesting information.

*Figure 8.* Appearance of the oil of slick C 2 h after release: the sea is very calm and there has been no formation of emulsion.

Figure 9 shows the slick B before and after the dispersant application; the relative areas of different colours (therefore different thicknesses) have been assessed. It can be observed that the dispersion was partial and some thick parts of the slick remained on the sea surface. Nonetheless, looking upwind these thick parts some orange colour can be seen which brings evidence that some dispersion occurred despite the very calm sea conditions. More, the evolution of the relative areas of different thicknesses (see the histograms on the right) shows a relative reduction of the thickest parts for the benefit of the thinnest parts.

Slick C was treated by the ship Lynx for 1 h and 10 min. The dispersant was applied neat with adjustable flow rate spraying equipment from the French Navy; this equipment is composed of 3 spraying assemblies, which can be operated alone or simultaneously to get different flow rates (between 10 and 90 L/min). According to the speed of the ship, the treatment rate can be adapted to the amount of oil to be treated (e.g., at 8 kt, from 20 to 150 L/ha). For the treatment, the Lynx cruised at 10 kt, decreasing to 8 kt at the end to treat the thickest part of the slick. Figure 10 shows slick C before and after the dispersant application. Similarly to slick B, a partial dispersion can be observed, with thick oil patches remaining on the sea surface but also a relative reduction of the thickest parts for the benefit of the thinnest ones. Figure 11, which is an image from Actimar of slick C, confirms that dispersion has been partial: some dispersed oil (yellow to pink plume), the track of remaining thick oil (red) and the ship administering treatment at the bottom of the picture can be seen.

A comparison between the two slicks B and C after treatment can be done on the aerial imagery (IR/UV) taken by the French Customs – see Figure 12; assessment of the IR picture shows a higher reduction of the thickest part of slick B than for slick C: slick B, thick area (white) 11 ha, medium thickness area (black) 8 ha. – slick C; thick area (white) 29 ha, medium thickness area (black) 55 ha. This observation is quite surprising because, with very calm weather, the shipborne treatment which brings extra mixing energy with the ship bow wave should have been more efficient; (such an observation had been made

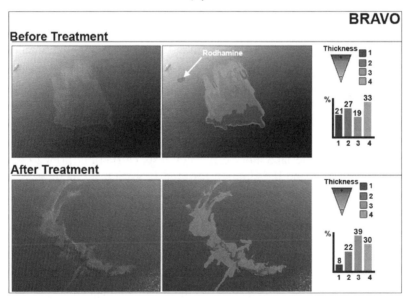

*Figure 9.* Aerial pictures visible spectrum, of slick B before and after dispersant treatment: left – aerial photography; right – the original pictures have been processed to assess the relative areas covered by the different oil thicknesses (red emulsion to sheen) – see the histogram.

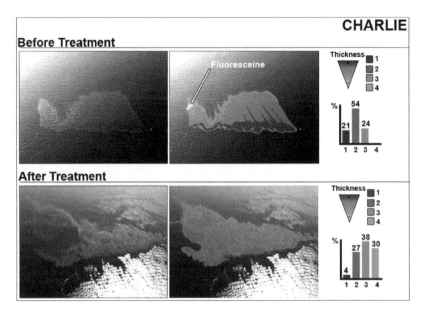

*Figure 10.* Aerial pictures visible spectrum, of slick C before and after dispersant treatment: left – aerial photography; right – the original pictures have been processed in order to assess the relative areas covered by the different oil thickness (red emulsion to sheen) – see the histogram.

during Protecmar 6 sea trials in 1986 (Bocard *et al.*, 1987): comparison of treatment in very calm weather with a ship and an helicopter). Therefore, the difference observed between slick C and B can likely be attributed to the dispersant used, the Finasol OSR 62 used on slick B seems to be more efficient than Gamlen OD 4000 used on slick C. This is reflective of the laboratory efficiency tests carried out prior to the experiment with the IFP dilution test: the FINASOL OSR 62 gave a slightly better efficiency (E = 80) than the Gamlen OD 4000 (E = 74).

Another possible explanation for this difference could be the targeting of the slick: the Cessna POD had been continuously and directly guided by the spotting British aircraft during the treatment while the ship was guided by a few smoke canisters launched by the French Navy helicopter to mark out the thickest parts (see Figure 13). In this respect, slick B may have received better treatment than slick C.

Traces of these slicks were detected the following day: this oil was spread out as a very thin layer which tended to break out and self disperse when subjected to some agitation; these traces were no longer

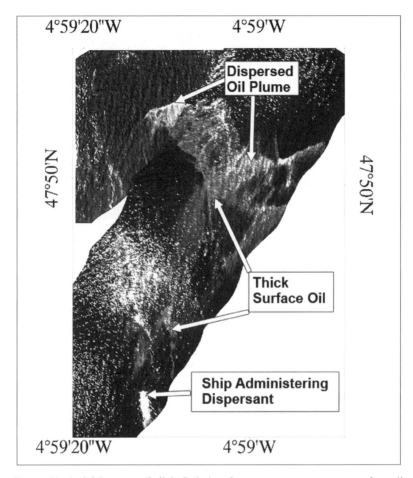

*Figure 11.* Aerial imagery of slick C during the treatment; we can see surface oil, plume of dispersed oil and the ship administering dispersant (Actimar).

detectable on the third day. Despite the fact that the dispersion process had obviously been limited due to the absence of natural agitation, the dispersion of the slick did occur with time.

## 4.  Depol 04 General Conclusions

The Depol 04 sea trials organized in May 2004 off the coast of Brittany by CEDRE, the French Navy and French Customs, were designed to study the natural weathering of paraffinic and asphaltenic oils and to

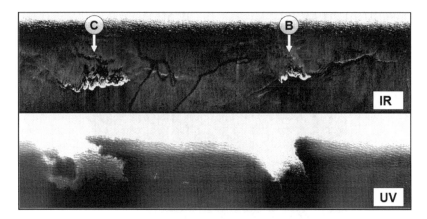

*Figure 12.* Comparison of slicks B and C on IR and UV imagery.

*Figure 13.* Treatment by the ship guided with smoke canisters and by the Cessna POD.

assess the efficiency of dispersant treatments. Aircraft Cessna POD spraying equipment from OSRL and shipborne spraying equipment from the French Navy were used to apply Finasol OSR52 and Gamlen OD 4000 dispersants, respectively. In addition, Depol 04 presented an opportunity to run the annual *Bonnex* intercalibration exercise of the remote sensing capabilities of the Bonn Agreement members and to test new remote sensing devices developed by Actimar and NMRI.

The meteorological conditions during these sea trials remained very calm (sea state mainly between 0 and 2) and was not very suitable for the dispersion process due to the lack of natural mixing energy. Nonetheless, on the first slick (paraffinic oil) significant dispersion occurred and led to a large reduction of the residual surface oil: after two days, the weathered oil emulsion recovered at the sea surface represented only 1 m$^3$ while 10 m$^3$ of pure oil had originally been released.

Despite the guidance on the slick provided by the UK spotter aircraft, the dispersant treatment carried out by the Cessna POD system did not succeed in covering the entire slick area; particularly, some of the thickest areas were not treated. This observation demonstrates the need to improve the procedures for guiding treatment vessels and subsequently optimizing the effectiveness of dispersant in the areas being treated.

The last two slicks (asphaltenic oils) were treated by the Cessna POD with Finasol OSR 62 and by shipborne adjustable spraying equipment for neat dispersants with Gamlen OD 4000, respectively. A short time after dispersant application, the aerial imagery of the treated slicks showed that some dispersion had occurred in both slicks despite the very calm weather (sea state 0–1); however, although partial, the dispersion of the slick treated by the Cessna POD appeared to be better than that of the slick treated by the ship. This observation indicates that the dispersant Finasol OSR 62 was more efficient than Gamlen OD 4000 and/or that the Cessna POD guided by the UK spotter aircraft targeted the slick better than the ship which was guided by smoke canisters launched by a helicopter on the thickest areas of the slick.

Both observations militate in favour of improving the operational procedures. During the last two decades dispersant formulations and spraying devices have been studied and improved, with the result that dispersion procedures are now the major limiting factor.

## 5.  Depol 05 Presentation

Depol 05 was a sea trial specifically devoted to assess at sea, the intrinsic efficiency of dispersants on viscous oil.

### 5.1. BACKGROUND

In 2003 OSRL, MCA and ITOPF joined in an effort to run sea trials devoted to assessing the dispersibility of viscous oils according their viscosity, the dispersant and the dispersant to oil ratio (DOR: 1/25, 1/50, 1/10). The trial consisted in spilling very small slicks (a few tens litres) of fuel oils IFO 180 and 380, which were subsequently treated with dispersant and visually assessed for dispersion efficiency by a panel of six experts according to four criteria:

0 = No dispersion
1 = Slow or partial dispersion
2 = Moderately rapid dispersion
3 = Very rapid and total dispersion

The oil was released and treated from a barge cruising at low speed, the experts were in a small boat looking carefully at the slick and noting their observations at 2, 5 and 10 min after the treatment.

Twenty-six tests were run over several days and their results led to interesting conclusions: while IFO 180 was easily dispersed, the dispersion of IFO380 was much more difficult; however, if IFO380 dispersibility was low, it was increasing with the dosage of dispersant. The remaining question was: what would occur with a higher dispersant dosage? Moreover, for each test the observations lasted only 10 min; another remaining question was: would dispersion occur over a longer period of time? In addition, due to the experimental design (i.e., spraying equipment) the dispersant dosage was poorly controlled, as a large portion of the dispersant was applied onto the water surface aside the oil and did not have sufficient time to spread across the slick.

*Figure 14.* UK sea trials; oil and dispersant applications from the barge, and the observers' boat.

## 5.2. DEPOL 05 OBJECTIVES AND PRINCIPLES

The Depol 05 experiment was designed to pursue and improve the work initiated in the UK. The objective was to see if viscous oils were amenable dispersal with higher dispersant dosage; the experimental plan considered four oil viscosities (2,000, 5,000, 8,000 and 10,000 cSt), four dispersant dosages (0%, 5%, 10% and 15%) and three dispersants. The principle of the experiment was quite similar to what had been done in the UK with the following improvements.

### 5.2.1. *Observation Criteria*

To get a better observation of dispersion, observers were requested to specify what they observed according to the following four criteria:

- Observation of a plume of dispersed oil
- Observation of oil resurfacing
- Observation of a spreading effect on the oil
- Observation of a white cloud indicating some dispersant being diluted directly in the sea.

### 5.2.2. *Oil Release and Dispersant Application Equipment*

A floating open corridor built with two parallel booms held at both ends by frames, was towed aside a ship cruising at slow speed. The oil was spilled from the first frame at the entrance of this corridor and the dispersant was applied on the oil from the second frame at the exit of this corridor. This was done to ensure that the oil had time to spread across and that all the oil would be treated with dispersant. The dispersant equipment was composed of three spraying booms which could be activated independently to get 5%, 10% or 15% dispersant dosage with the oil.

*Size of the slicks*:
The volume of each oil slick was 150 L in order to have enough time to observe the slick for half an hour. The oil for each slick was prepared from a mixture of heavy fuel and kerosene to obtain the requested viscosity.

*Figure 15.* The "corridor" system.

*Figure 16.* Oil released from the first frame at the entrance of the corridor and dispersant application from the last frame and the exit of the corridor.

### 5.2.3. *Progress of the Sea Trials*

Unfortunately due to a improper manoeuvre, the corridor system was destroyed by the propeller of the ship during the third test: only 2 tests were carried out of the 22 tests initially planned in the experimental program; therefore, it was not possible to assess the limit of dispersibility for oil in relation to the dispersant type and dispersant dosage. Nonetheless, the two tests that were performed demonstrated the corridor system suitability to apply the oil and dispersant in a well controlled manner. These two tests also demonstrated the importance of the observers' location in regards to the slick: observers should remain close to the slick and without disturbance due to the sun reflection.

## 5.3. DEPOL 05 CONCLUSIONS

The testing methodology used in DEPOL 05 proved to be promising to conduct repeatable and reliable tests, providing the observers are suitably located close enough to the slick to be observed. Using small volume slicks, it is possible to complete a much larger number of tests considered in an experimental matrix and therefore answer question such as oil dispersibility, dispersant efficacy, etc....

## 6. General Conclusions

The sea trials program developed in France regarding chemical dispersion involved two types of testing methodology:

- Large sea trials were designed for large scale oil slicks and involved operational means (i.e., planes and ships) to globally test dispersion techniques in realistic conditions. As main conclusions, these trials demonstrated: that dispersion could occur, at least partially, in rather calm sea conditions; and, pointed out operational limitations such as the difficulty to target and treat the entirety of the thickest part of a slick.

- Small scale trials were designed to control for, as much as possible, the testing conditions (oil spillage and dispersant application), the intrinsic efficiency of dispersant products, and to specify the oil viscosity limits for dispersion. Unfortunately, due to a navigation failure these tests were interrupted without giving clear answer to the question of dispersibility. Nonetheless, the testing procedure proved to be well adapted to carrying out such tests, and we hope to resume this experimental program in the near future.

## 7. Acknowledgements

This experiment would not have been possible without the contribution of Captain Nedelec, head of the antipollution technical team of the French Navy, (CEPPOL), Captain Le Nouy, on scene commander during the experiment, and Mr Castanier, officer in the French Customs. Thanks should be expressed to the other participants, including Ms Dimercantonio from MUMM, Mr Melbye from SINTEF, Mrs Varescon and Lavigne from TOTAL S.A, M Grenon from ITOPF, M Lewis, the OSRL operators, the crews of the Bonn Agreement remote sensing aircrafts involved, and the laboratory team of CEDRE.

## References

Bocard, C., Castaing, G., Ducreux, J., Gatellier, C., Croquette, J. and Merlin, F., 1987, Protecmar: The French Experience from a Seven-year Dispersant Offshore Trials Program: International Oil Spill Conference, Baltimore, Maryland.

Guyomarch J., Mamaca, E., Champs, M. and Merlin, F.-X., 2002, Oil Weathering and Dispersibility Studies: Laboratory, Flume, Mesocosms and Field Experiments, in: Proceedings of the 3rd IMO R&D Forum.

Guyomarch J., Morin, E., Goutard, A. and Merlin, F.-X., 2001, Experimental Oil Weathering Studies in Hydraulic Canal and Open Pool to Predict Oils Behaviour in Case of Casual Spillage, in: Proceedings of the 2001 International Oil Spill Conference, American Petroleum Institute, Washington, DC.

# DISPERSANT RESEARCH IN A SPECIALIZED WAVE TANK: MIMICKING THE MIXING ENERGY OF NATURAL SEA STATES

A.D. VENOSA[†]
*U.S. Environmental Protection Agency Cincinnati, OH 45268, USA*

K. LEE & Z. LI
*Fisheries and Oceans Canada, Dartmouth, Nova Scotia B2Y 4A2, Canada*

M.C. BOUFADEL
*Temple University, Philadelphia, PA 19122, USA*

**Abstract[*].** Breaking waves play a crucial role in the dispersion of oil spilled on the surface of the ocean both in the presence and absence of oil dispersants. Breaking of waves occurs when the forward horizontal velocity of water in the wave crest is greater than the wave propagation speed. These waves cause velocity shear and hence result in the mixing of oil and dispersant. Velocity shear with its associated friction also causes the dissipation of kinetic energy of the fluid. Of interest is the kinetic energy dissipation rate per unit mass, $\varepsilon$, which varies both in time and space. We use velocity measurements to compute the shear and subsequently the energy dissipation rate. The effectiveness of chemical oil dispersants is typically evaluated at various wavelength scales ranging from the smallest (10 cm, typical of laboratory flasks) to the largest (10s to 100s m, typical of open ocean test conditions). This study aims at evaluating dispersant effectiveness at intermediate or pilot scale. The hypothesis is that the energy dissipation rate per unit mass, $\varepsilon$, plays a major role in the effectiveness of a dispersant. If one assumes that $\varepsilon$ adequately describes dispersion behavior at both wave tank and field scales, then differences in the overall effectiveness of a dispersant under field conditions should be resolvable. To quanti-

---

[†] To whom correspondence should be addressed.  E-mail: venosa.albert@epa.gov
[*] Full presentation available in PDF format on CD insert.

W. F. Davidson, K. Lee and A. Cogswell (eds.), *Oil Spill Response: A Global Perspective.*   141
© Springer Science + Business Media B.V. 2008

tatively define the conditions needed for effective dispersion in the field using scientifically sound, reproducible techniques, a wave tank measuring 32 m long × 0.6 m wide × 2 m deep was constructed on the premises of the Bedford Institute of Oceanography, Dartmouth, Nova Scotia. Waves are generated using a flap-type wavemaker. Controlled breaking wave conditions are generated by operating the wave maker at a low frequency followed by a higher frequency. Experiments defining the velocity profile and energy dissipation rates in the wave tank were conducted at three different induced breaking-wave energies. Energy dissipation rates were measured with an Acoustic Doppler Velocimeter (ADV) coupled to a data acquisition system. This presentation summarizes those results as well as preliminary results of the first series of dispersant effectiveness experiments showing the important influence breaking waves have on oil droplet dispersion.

**Keywords:** dispersants, wave tank, oil spill response, waves, oil droplets

# WAVE TANK STUDIES ON CHEMICAL DISPERSANT EFFECTIVENESS: DISPERSED OIL DROPLET SIZE DISTRIBUTION

Z. LI[†], K. LEE, P. KEPKAY, T. KING & W. YEUNG
*Center for Offshore Oil and Gas Environmental Research, Fisheries and Oceans, Canada, Dartmouth, Nova Scotia, B2Y 4A2, Canada*

M.C. BOUFADEL
*Department of Civil and Environmental Engineering, Temple University, Philadelphia, Pennsylvania, 19122, USA*

A.D. VENOSA
*National Risk Management Research Lab, U.S. EPA, Cincinnati, OH 45268, USA*

**Abstract.** In evaluation of chemical dispersant effectiveness, the two most important factors that need to be addressed and fully characterized in terms of efficacy are energy dissipation rate and particle size distribution. A wave tank facility was designed and constructed to specifically address these factors in controlled oil dispersion studies. The particle size distribution of the dispersed oil was quantified by a laser *in-situ* scattering and transmissometer (LISST-100X). The size distribution and morphology of the dispersed oil were characterized by an image analysis system based on a microscope fitted with transmitted light and ultraviolet-epifluorescence illumination. Time-series particle size distribution during physical and chemical dispersion of crude oil under a variety of non-breaking and breaking waves are presented.

**Keywords:** wave tank, dispersant, oil droplet, LISST, waves

---

[†] To whom correspondence should be addressed. E-mail: LiZ@mar.dfo-mpo.gc.ca

W. F. Davidson, K. Lee and A. Cogswell (eds.), *Oil Spill Response: A Global Perspective.* 143
© Springer Science + Business Media B.V. 2008

## 1. Introduction

The application of chemical dispersants is considered to be one of the primary oil spill countermeasures for reducing the overall environmental impact of marine oil spills (NRCNA, 2005; NRC, 2005; Lessard and Demarco, 2000). In addition to operational convenience, application of dispersants to treat oil slicks on the sea surface has advantages to minimize the harmful effect of floating oil on animals such as birds and marine mammals that frequent the water surface, and to reduce the risk of oil slicks contaminating coastal and/or shore-line environments.

Dispersants are chemicals that contain surfactants that reduce the surface tension between oil and water, resulting in the formation of oil droplets (oil-in-water emulsion). The dispersion of oil slicks is significantly enhanced in the presence of waves. Waves provide mixing energy, which breaks the surface oil film and propels oil droplets into the water column. Thus, in the context of oil spill response operations, dispersion is a physical-chemical process, whose effectiveness depends on the chemical properties of both dispersant and the oil and the mixing energy generated by the physical action of waves (NRC, 2005; Fingas, 2000). The hydrodynamic behaviour may dramatically influence natural and chemical dispersion of oil (Shaw, 2003; Delvigne and Sweeney, 1988). In particular, breaking waves play a crucial role in the mixing of oil and dispersant and hence the dispersion of an oil slick (Shaw, 2003). Breaking of waves occurs when the forward horizontal velocity of water in a wave crest is greater than the wave propagation speed. These waves cause velocity shear and hence result in the mixing of oil and dispersant. In turbulent flows, the velocity shear results from both spatial and temporal (turbulent) variation of velocities, but usually the turbulence contribution is dominant. Velocity shear with its associated friction also causes the dissipation of kinetic energy of the fluid. Of interest is the kinetic energy dissipation rate per unit mass, $\varepsilon$, which varies both in time and space. One may use velocity measurements in a selected water body to compute the shear, and subsequently the energy dissipation rate.

The effectiveness of a particular dispersant is typically evaluated at various scales ranging from the smallest (10 cm, typical of the Swirling Flask Test in the laboratory) to the largest (10s to 100s m, typical of field scale open water dispersion tests). In terms of product selection for spill response operations, standard laboratory assays for the evaluation of oil dispersant effectiveness such as the swirling flask test have limitations due to insufficient mixing energy and/or failure to account for the transport and interaction between oil and dispersant in water column (Sorial, 2004a, b; Venosa *et al.*, 2003). Testing at sea, however, is expensive and not always reproducible due to uncontrolled environmental variables, and hence unrealistic for routine testing of different dispersants on different oils. To address these concerns, a wave tank facility was constructed for evaluation of chemical oil dispersant effectiveness at intermediate or pilot scales.

The current hypothesis is that the energy dissipation rate per unit mass, $\varepsilon$, plays a major role in the effectiveness of a dispersant. Conservation of $\varepsilon$ between the wave tank and actual field conditions provides support for the use of our test system to evaluate the operational effectiveness of chemical oil dispersants. Preliminary hydrodynamic tests have demonstrated that the non-breaking waves and breaking waves that were generated in our test tank facility were similar to the reported energy dissipation rates for natural waters (Venosa *et al.*, 2005; Delvigne, 1988).

## 2.  Materials and Methods

### 2.1.  WAVE TANK FACILITIES

Figure 1 presents the schematic of the wave tank. The tank facility measures 32 m long, 0.6 m wide, and 2 m high. The water depth during the present experiments was 1.50 m. Different waves can be generated by a paddle situated at one end of the tank linked to an adjustable cam that controls its stroke length to alter wave-height characteristics. The wave frequency (and subsequently wave length) is controlled by the rotation speed of the cam. The computer-controlled wave-generator is capable of producing both regular non-breaking waves and breaking waves with designated length, height and frequency. The system is very useful for dispersion studies because recurrent breaking of waves can be generated at the same location. This is done by superimposing a wave of one frequency onto a wave of another frequency, causing the wave to break under different inertial forces. Calibration of non-breaking

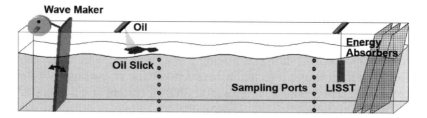

*Figure 1.* Schematic representation of the wave tank.

and breaking-wave energy was conducted using a scalable parameter, energy dissipation rate. The details of wave energy calibration have been reported elsewhere (Venosa *et al.*, 2005).

## 2.2. EXPERIMENTAL PROCEDURES

The wave tank study on chemical dispersant effectiveness testing was conducted using two crude oils (MESA and Alaska North Slope (ANS) oil), three dispersants (water, Corexit 9500 and SPC1000), and three wave energy conditions (a regular wave, a spilling breaking wave, and a plunging breaking wave).

A three-factor mixed-level factorial experiment was designed. The three factors and their levels are: two oils, three dispersants, and three waves. Hence, 18 treatments were set up for this dispersant effectiveness study experiment; with triplicate runs for each treatment, resulting in 54 runs for the entire experiment. Different treatments were conducted in a random order to minimize the impacts of other confounding factors such as temperature, salinity, and wind on the dispersant effectiveness of crude oil.

## 2.3. TOTAL PETROLEUM HYDROCARBONS

The dispersed oil in aqueous samples was extracted with dichloromethane and measured with a DU series 60 ultraviolet-visible spectrophotometer (Beckman Instruments, Inc., Fullerton, CA) capable of measuring absorbance at 340, 370, and 400 nm (Venosa *et al.*, 2002). These absorbance values were used because they represent different locations within the absorbance curve. A 10 mm cuvette with 1 cm path length was used for measurement and dichloromethane was used as the reference blank. For all analyses the cuvette was used with a polytetrafluoroethylene cover.

The direct ultraviolet fluorescence spectroscopy was also applied to monitor the dispersed/dissolved oil in seawater using a method reported previously (Kepkay *et al.*, 2002). Briefly, samples collected from the wave tank at specified times and locations were vigorously shaken by hand, and 3 mL of the suspension was rapidly transferred to an ultraviolet-grade methyl acrylate disposable cuvette (VWR International Inc., Mississauga, ON). The suspension was immediately scanned in the dissolved/dispersed fraction using a QM-1 spectrofluorometer running FeliX software (PTI, Inc., Birmingham, NJ). The optimal excitation wavelength that produced the highest emission peaks was 320 nm. This wavelength with a slit width of ±2 nm was used in all subsequent emission scans from 340 to 500 nm.

## 2.4. PARTICLE SIZE DISTRIBUTION

Oil droplet size distribution inside the wave tank was determined by a Type C LISST-100X particle counter (Sequoia, Seattle, WA), which has 32 particle size intervals logarithmically spaced from 2.5 to 500 mm in diameter, with the upper size in each bin 1.18 times the lower. Particle size distribution is expressed as the average volumetric concentration of oil droplets falling into each interval of the size range. In general, the particle size distribution measured using LISST fits a lognormal distribution, which has been extensively used for measuring aerosol size distribution in natural environment (Hinds, 1999). The data acquisition is conducted at real time operation mode throughout each experimental run, with an average of ten measurements for each sample being taken every 3 s. The *in-situ* dispersed oil droplet size distribution and total oil concentration are measured at one horizontal location (16 m downstream of the flap) and three different depths (30, 75, and 120 cm under water) over eight continual time periods.

## 3.  Results and Discussion

### 3.1. HYDRODYNAMICS OF THE WAVE TANK

Three different wave conditions were selected to represent the typical wave energy conditions at sea. Photos of the three wave conditions are presented in Figure 2. The three wave types are: (1) regularly non-breaking waves; (2) spilling breakers; and (3) plunging breakers.

Regular non-breaking and breaking wave profiles were recorded using wave gauges that were deployed at different locations of the wave tank. For the regular non-breaking waves, uniform waves were produced throughout the entire length of the wave tank, including the initial wave generation zone, intermediate wave propagation zone, and the end of the wave tank nearest to the wave absorbers. The observed wave height (0.17 m) matches well with the theoretical prediction based on linear theory of waves (0.16 m). The constant wave heights throughout the tank suggest that the energy dissipation rate of the regular waves was small. Most of the wave energy was absorbed when wave were propagated to the end of the tank by the wave absorbers. The smooth wave profiles recorded by the wave gauge near the end of the wave tank indicates the reflection of the tail end of the tank was effectively controlled through the proper functioning of the wave absorbers.

Breaking waves were generated by superposition of two different waves. The high frequency with high wave height (0.17 m) followed by low frequency with low wave height (0.11 m), is clearly shown with the wave profile obtained at 2 m from the wave generation flap. Superposition of the waves occurred immediately prior to the breaking zone, creating a transitional wave height as much as 0.28 m, which is equal to the sum of the two component waves. The instant breaking waves and dissipation of wave energy in the mixing zone were captured by the wave gauges that are deployed in the breaking zone and the one further down stream of the wave tank. The post breaker wave height was reduced to 0–0.03 m, indicating that most of the wave energy was dissipated from the surface water body to the bulk water body as micro-scale turbulence.

Similar wave profiles were obtained for the spilling breaking waves, one with a stroke of 6 cm and the other with a stroke of 7 cm. The observed maximum wave height (after superposition of the two components) was 0.21 and 0.25 m, respectively, for S = 6 and 7 cm conditions. Breaking waves occurred at approximately the same location as the plunging breaker. The spilling breakers, especially for S = 6 cm, were less violent than the plunging breaker. Since wave energies are proportional to the square of wave height, spilling breakers are likely to have low energy dissipation rates compared to plunging

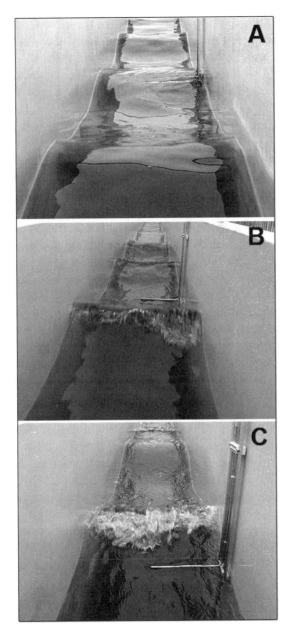

*Figure 2.* Photographs of three representative wave conditions: (A) regular non-breaking wave, (B) spilling breaking wave, and (C) plunging breaking wave.

breaking waves, assuming the time scale of these breakers are similar. Given our goal is to find a variety of representative wave conditions of the real sea states, the spilling breaker with a stroke of 7 cm was selected along with the plunging breaker and the regular non-breaking wave for the next chemical dispersant effectiveness test.

## 3.2. DISPERSANT EFFECTIVENESS STUDY

A preliminary dispersant effectiveness study was conducted in the wave tank under three wave conditions. Dispersant effectiveness was evaluated by monitoring oil distribution using the LISST-100X and ultraviolet fluoremeter (UVF).

Figure 3 shows the total dispersed oil concentration at three different depths, namely near the surface, in the middle, and near the bottom, as recorded by LISST-100X. Oil concentrations at the surface and in the middle of the tank are similar, but they appear to be more dynamic at the bottom. The fluctuation of dispersed oil concentration with time was pronounced during the first hour under a breaking wave regime. After the paddle was stopped to maintain the tank in a quiescent state, oil concentrations were less variable for all three depths. The total oil concentrations slightly increased over time, indicating that resurfacing of the dispersed oil was effective at quiescent conditions. The total oil concentration distribution in the wave tank over time was consistent with the dispersed oil droplet size distribution.

Figure 4 shows the mass mean diameter (MMD) of the dispersed oil droplet size distribution at three different depths in the water column. At near surface and in the middle of the tank water, oil droplets were usually less than 100 μm, whereas at near bottom, MMD of the dispersed oil droplets were more than 200 μm within the first hour and declined to less than 100 μm after 2 h dispersion. The MMD of the dispersed oil droplets remained constant throughout the last 2 h during the quiescent hydrodynamic regime, indicating that although resurfacing the submerged oil droplets is inevitable, recoalscence of the small dispersed oil droplets into large oil droplets may not necessarily occur because of the presence residual surfactant from the chemical dispersant.

Total oil concentration and dispersed oil droplet size distribution data were also obtained for the regular wave and the spilling breaking

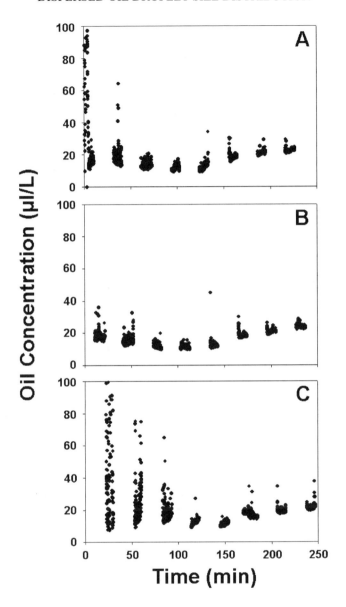

*Figure 3.* The total dispersed oil concentration measured using LISST-100X: (A) near surface (45 cm), (B) in the middle (80 cm), and (C) near bottom (120 cm).

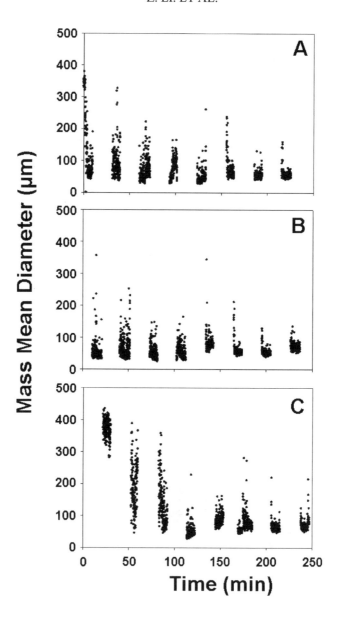

*Figure 4.* The mass mean diameter of the dispersed oil droplets measured using
LISST-100X: (A) near surface (45 cm), (B) in the middle (80 cm), and (C) near bottom
(120 cm).

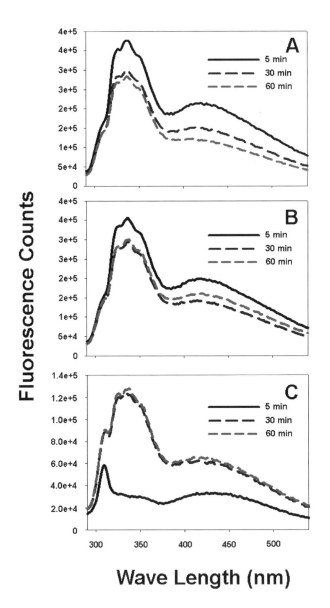

*Figure 5.* The fluorescence response (UVF) of samples taken from downstream end of the wave tank (L = 20 m from the flap) from three depths: (A) near surface (5 cm), (B) in the middle (75 cm), and (C) near bottom (140 cm).

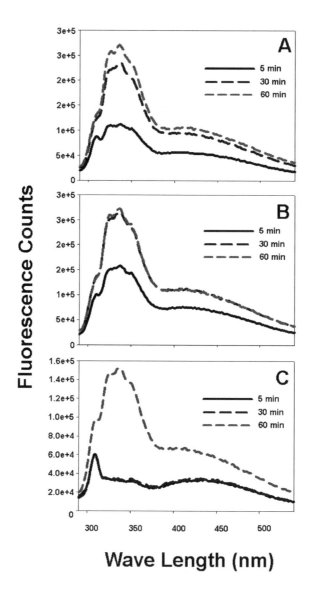

*Figure 6.* The fluorescence response (UVF) of samples taken from downstream middle of the wave tank (L = 14 m from the flap) from three depths: (A) near surface (5 cm), (B) in the middle (75 cm), and (C) near bottom (140 cm).

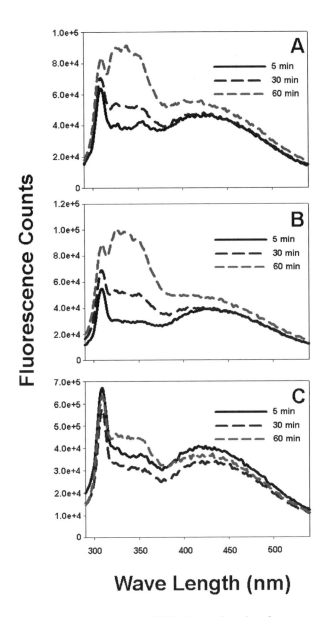

*Figure 7.* The fluorescence response (UVF) of samples taken from upstream of the wave tank (L = 8 m from the flap) from three depths: (A) near surface (5 cm), (B) in the middle (75 cm), and (C) near bottom (140 cm).

wave conditions. Generally the total oil concentrations were low near the bottom for these wave conditions, and the observed dispersed oil droplet sizes were also much smaller in this area of the water column.

Figures 5–7 present the emission spectra of the crude oil dispersed at three different horizontal locations (upstream, near downstream, and further downstream of the oil addition spot), three depths (near surface, middle, and near bottom), and over time (5, 30, and 60 min after wave generation). The initial transport of oil is fast, so that the fluorescence counts were the highest at the surface and middle of the water column at the further downstream location (Figure 5A, B). UVF signals also dramatically increased at near bottom at this location with progress of oil dispersion (Figure 5C).

At near downstream location, UVF signals all increased with the progress of dispersion, suggesting that there was a back flow of the oil mass at the bottom of the wave tank after the initial rapid transport to further downstream (Figure 6).

The back flow of oil mass was more clearly demonstrated by the UVF signals shown in the samples taken from the upstream location (Figure 8A, B). The marked increase of the UVF fluorescence counts near surface and in the middle of the water column indicated that dispersed oil droplets were transported upstream by the under water currents. However, the relatively weak signals at near bottom of the upstream sampling location suggest that the dispersed oil droplets may be moving upwards in the absence of turbulence that was produced by energy dissipation from breaking waves in this upstream area of the wave tank (Figure 7C).

## 4. Conclusions

The data reported in this paper support the following conclusions: First, oil dispersion effectiveness was correlated with energy dissipation rate; elevated dissipation energy promotes the penetration of oil into the bulk aqueous phase; and the presence of dispersant increased the dispersed oil concentration at the same energy levels. Second, the chemical dispersant significantly reduced the oil droplet sizes, especially at low energy states. Third, re-surfacing of oil occurred at static conditions; and the stability of dispersed oil is significantly increased in the presence of dispersant.

## 5. Acknowledgements

This research was supported by the Panel of Energy Research and Development (PERD) Canada, U.S. EPA (research contract No. 68-C-00-159), and NOAA/UNH Coastal Response Research Center (Grant Number: NA04NOS4190063 UNH Agreement No.: 06-085). Essential technical and logistical support to this research program was provided by Susan Cobanli, Jennifer Dixon, Xiaowei Ma, Peter Thamer, and Matt Arsenault.

## References

Delvigne, G.A.L. and Sweeney, C.E., 1988, Natural dispersion of oil. *Oil and Chemical Pollution*, **4**(4), 281–310.

Fingas, M.F., 2000, Use of Surfactants for Environmental Applications. *In Surfactants: Fundamentals and Applications to the Petroleum Industry*, Schramm, L.L., Ed. Cambridge University Press, Cambridge, pp. 461–539.

Hinds, W.C., *Aerosol Technology: Properties, Behavior, and Measurement of Airborne Particles*, 2nd ed. Wiley, New York, 1999.

Kepkay, P.E., Bugden, J.B.C., Lee, K., and Stoffyn-Egli, P., 2002, Application of ultraviolet fluorescence spectroscopy to monitor oil-mineral aggregate formation. *Spill Science and Technology Bulletin*, **8**(1), 101–108.

Lessard, R.R. and Demarco, G., 2000, The significance of oil spill dispersants. *Spill Science and Technology Bulletin*, **6**(1), 59–68.

NRCNA, *Understanding Oil Spill Dispersants: Efficacy and Effects.* The National Academies Press, Washington, DC, 2005.

NRC, *National Research Council: Understanding Oil Spill Dispersants: Efficacy and Effects.* National Academies Press: Washington, DC, 2005.

Shaw, J.M., 2003, A microscopic view of oil slick break-up and emulsion formation in breaking waves. *Spill Science and Technology Bulletin*, **8**(5–6), 491–501.

Sorial, G.A., Venosa, A.D., Koran, K.M., Holder, E., and King, D.W., 2004a, Oil spill dispersant effectiveness protocol. I: impact of operational variables. *Journal of Environmental Engineering-Asce,* **130**(10), 1073–1084.

Sorial, G.A., Venosa, A.D., Koran, K.M., Holder, E., and King, D.W., 2004b, Oil spill dispersant effectiveness protocol. II: performance of revised protocol. *Journal of Environmental Engineering-ASCE,* **130**(10), 1085–1093.

Venosa, A.D., King, D.W., and Sorial, G.A., 2002, The baffled flask test for dispersant effectiveness: a round robin evaluation of reproducibility and repeatability. *Spill Science and Technology Bulletin*, **7**(5–6), 299–308.

Venosa, A.D., Kaku, V.J., Boufadel, M.C., and Lee, K., 2005, In Measuring energy dissi-pation rates in a wave tank. *International Oil Spill Conference*, Miami, FL, 2005, American Petroleum Institute, Washington DC, Miami, FL.

# WAVE TANK STUDIES ON FORMATION AND TRANSPORT OF OMA FROM THE CHEMICALLY DISPERSED OIL

K. LEE[†], Z. LI, T. KING & P. KEPKAY
*Center for Offshore Oil and Gas Environmental*
*Research, Fisheries and Oceans Canada, Dartmouth,*
*Nova Scotia, B2Y 4A2, Canada*


M.C. BOUFADEL
*Department of Civil and Environmental Engineering,*
*Temple University, Philadelphia, PA, 19122, USA*


A.D. VENOSA
*National Risk Management Research Lab, U.S. EPA,*
*Cincinnati, OH 45268, USA*

**Abstract.** The interaction of chemical dispersants and suspended sediments with crude oil influences the fate and transport of oil spills in coastal waters. A wave tank study was conducted to investigate the effects of chemical dispersants and mineral fines on dispersion of oil, formation of oil-mineral-aggregates (OMAs), and microbial activities in natural seawater. Results of ultraviolet fluoremetry (UVF) and gas chromatography-flame ionized detector (GC-FID) analysis indicate that both dispersants and mineral fines, alone and in combination, stimulate the dispersion of oil slick from surface to water column. A laser *in-situ* scattering and transsiometer (LISST-100X) measurement shows that the presence of mineral fines increased the total concentration of the suspended particles from 4 to 10 µL/L, whereas the presence of dispersants decreased the particle size (mass mean diameter) from 50–70 to 20 µm. Enumeration with epifluorescent microscope shows that the presence of either dispersants or mineral fines significantly increased the number of particles in water.

**Keywords**: wave tank, dispersant, OMA, LISST, fluoremetry

---

[†] To whom correspondence should be addressed. E-mail: LeeK@mar.dfo-mpo.gc.ca

W. F. Davidson, K. Lee and A. Cogswell (eds.), *Oil Spill Response: A Global Perspective.*    159
© Springer Science + Business Media B.V. 2008

## 1. Introduction

Accidental release of crude oils in coastal waters is relatively common due to escalated offshore oil and gas exploration and production activities, petroleum pipeline and tanker transport, and natural seeps (Anderson and LaBelle, 2000; NRC, 2003). Natural dispersion of crude oils by wave actions leads to formation of micron-sized droplets that are eventually diluted through mixing processes in the sea (Delvigne and Sweeney, 1988; Li and Garrett, 1998; Tkalich and Chan, 2002; Shaw, 2003). In nearshore or estuarine waters, oil droplets are incurporated into oil-mineral-aggregates (OMAs) as the droplets interact with the high loads of suspended particulates that are typical in coastal regions (Bragg and Owen, 1995; Page et al., 2000; Le Floch et al., 2002; Owens and Lee, 2003; Owens et al., 2003). Detailed studies of OMA formation have revealed that both mineral fines and organic particles can stabilize oil droplets within the water column (Delvigne et al., 1987; Bragg and Yang, 1995; Lee, Weise et al., 1996; Lee and Stoffyn-Egli, 2001; Lee, 2002; Muschenheim and Lee, 2002; Omotoso et al., 2002; Lee et al., 2003). The results from both controlled laboratory experiments (Lee et al., 1997; Cloutier et al., 2002; Omotoso et al., 2002; Stoffyn-Egli and Lee, 2002) and shoreline field trials (Owens et al., 1995; Lee et al., 1997; Lunel et al., 1997; Owens and Lee, 2003) demonstrated that the production of OMAs enhances the natural dispersion of oil spills and reduces their environmental persistence.

Presently, the utilization of dispersants in coastal regions has been approached with caution due to concerns over the potential exposure of benthic and aquatic organisms to the chemically-dispersed oil. Conflicting results have been reported in regard to the effects of dispersants on formation and fate of OMAs in both controlled studies and field observations (MacKay and Hussain, 1982; Gearing and Gearing, 1983; Guyomarch et al., 1999; Page et al., 2000; Guyomarch et al., 2002). For example, whereas MacKay and Hussain (1982) found that chemically-dispersed oil associated less with mineral fines than naturally dispersed oil, and Gearing and Gearing (1983) reported that the presence of high dose of suspended particles had little effect on removal of chemically-dispersed oil in water column, Guyomarch et al. (1999, 2002) did measure high amount of oil being incorporated in OMAs when various oils and chemical dispersants were mixed with clay minerals in shaking flasks. Theoretically the application of dispersants may change the performance of oil-mineral aggregation compared to the untreated oil. One argument is that the alteration of droplet size by chemical dispersants may lead to formation of denser small particles versus the formation of OMAs of neutral or slightly negative buoyancy. The other contention is that chemically, dispersants may change the surface physicochemical properties of oil droplets and therefore impair the bounding forces of the oil to mineral fines. Hence a study to compare formation and fate of physically and chemically dispersed oil droplets and OMAs is important for us to gain a better understanding of the role of chemical dispersants on crude oil dispersion in suspended-particle-loaded nearshore and estuarine waters. The objectives of this study are: (1) to investigate the aggregation of mineral fines with physically or chemically dispersed oil, (2) to determine the dynamic particle-size distributions of the physically- and chemically-dispersed oil and OMAs, and (3) to study the subsequent transport of oil droplets or OMAs in the wave tank.

## 2.  Materials and Methods

### 2.1.  WAVE TANK FACILITIES AND GENERATION OF BREAKING WAVES

Figure 1 presents the wave tank facility [16 × 0.6 m (wide) × 2 m (high); average water level 1.25 m] that is used in oil-mineral aggregation experiments. Different wave patterns are generated by a paddle situated at the front of the tank. The paddle is linked to an adjustable cam that controls stroke length to alter wave-height. The wave frequency (and thus the wavelength) is controlled by the rotation speed of the cam. The system is unique in that recurrent breaking waves can be generated at the same location and oil dispersion can be followed once it is added at the same location. In this study, the wave tank was operated under breaking wave conditions to represent a proper sea state for dispersant application (Venosa et al., 2005). The stroke of the paddle was set at 7 cm. In each cycle, high frequency waves (1.0 Hz, wave height about 0.12 m) were produced for 20 s, followed by low frequency waves (0.4 Hz, wave height about 0.05 m) for 20 s. This allows the second string of long-period waves to catch up with the first string of short-period waves to create recurrent breaking waves at approximately 7.5 m downstream of the paddle. The energy dissipation rate of kinetic energy per unit mass of water is an important parameter in evaluating the extent of mixing and dispersion of oils at sea (Delvigne and Sweeney, 1988). In this study, energy dissipation rate was estimated using time-series of water velocity measurements at various depths and locations in tank. The energy dissipation rates of the wave tank under breaking wave conditions were in a range of 0.1–0.5 m2/s3, which are close to the upper limit of ocean surface layer and lower limit of breaking waves at sea (Delvigne and Sweeney, 1988). Details of the calibration of wave energy are reported elsewhere (Venosa et al., 2005; Boufadel et al., in preparation).

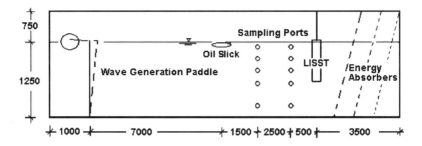

*Figure 1.* Schematic representation of the wave tank.

## 2.2. EXPERIMENTAL PROCEDURES

A two-factor two-level [dispersant (0, 3 mL) and mineral fines (0, 50 g)] factorial experimental design was used to investigate the effects of dispersants on formation and fate of OMAs. The test oil was 150 mL weathered Scotia Shelf Condensate, a light crude oil. The dispersant was Corexit 9500 (Nalco Energy Service, L.P. Sugar Land, TX) at a dispersant to oil ratio of 1:50. The mineral fines were Kaolinate (API # 9, Mesa Alta, New Mexico), which has a cation exchange capacity of 6.8 meg/110 g, a median particle size of 0.6 μm, and a density of 2.60 g/cm$^3$. The wave tank was filled with filtered natural seawater

(5 μm) from the Bedford Basin of Halifax harbour, with average water temperature 5°C and salinity 30 PSU. For each experimental run, 1 L premix (at a reciprocating shaker) of seawater + oil ± dispersant ± mineral fines was released on the surface in the middle of the tank prior to an incoming breaking wave. Samples were taken from the tank at two horizontal locations (1.5 and 4 m downstream of oil application), four depths (5, 20, 60, and 110 cm under water), and at five time points (1, 10, 30, 60, and 300 min after oil addition). The distribution of oil was determined by: (1) ultraviolet fluorescence (UVF) spectroscopy, and (2) solvent extraction and gas chromatography flame ionized detection (GC-FID) analysis of total petroleum hydrocarbon (TPH). The dynamic particle size distribution and total suspended particle concentration were measured using a laser *in-situ* scattering and transmissometer (LISST-100X). The morphology of the formed oil droplets and OMAs were observed using an epi-fluorescence microscope.

## 2.3. TOTAL PETROLEUM HYDROCARBONS

Samples collected from the tank were first filtered with micro-fiberglass filters (Whatman GF/C, Fisher Scientific, Canada) to separate the dissolved from the particulate phase. The liquid phase was transferred to a separatory funnel for extraction of hydrocarbon with dichloromethane using EPA method 3510C (EPA, 2003). The solid phase was processed using soxhlet extraction with dichloromethane over an 18 h cycle using EPA method 3540C (EPA, 2003). The extracts from both phases were concentrated to 1 mL and analyzed by GC-FID to measure TPH in the dissolved (milligram TPH per liter of tank water, or $mg \cdot L^{-1}$) and particle phase (mg TPH per gram of solids, or $mg \cdot g^{-1}$). The ratio of these concentrations ($L \cdot g^{-1}$) reflects the extent of partitionning of TPH between solid and liquid phases. Oil distribution as TPH in bulk aqueous phase as a summary of dissolved or aggregate phases inside the tank was also presented.

The direct ultraviolet fluorescence spectroscopy was also applied to monitor the dispersed/dissolved oil in seawater using a method reported previously (Kepkay *et al.*, 2002). Briefly, samples collected from the wave tank at specified times and locations were vigorously shaken by hand, and 3 mL of the suspension was rapidly transferred to an ultraviolet-grade methyl acrylate disposable cuvette (VWR International

Inc., Mississauga, ON). The suspension was immediately scanned in the dissolved/dispersed fraction using a QM-1 spectrofluorometer running FeliX software (PTI, Inc., Birmingham, NJ). The optimal excitation wavelength that produced the highest emission peaks was 320 nm. This wavelength with a slit width of ±2 nm was used in all subsequent emission scans from 340 to 500 nm.

### 2.4. PARTICLE SIZE DISTRIBUTION

Oil droplet size distribution inside the wave tank was determined by a Type C laser *in-situ* scattering and transmissometer (LISST-100X; Sequoia, Seattle, WA). There are 32 particle size intervals logarithmmically placed from 2.50 to 500 μm in diameter, with the upper size in each bin 1.18 times the lower. Particle size distribution is expressed as the average volumetric concentration of particles falling into each interval of the size range. In general, the particle size distribution from LISST fits a lognormal distribution, and the mass (or volume) mean diameter of the dispersed oil droplets is used to quantitatively compare the average size of the particles. The LISST was situated 4.5 m downstream from the oil application in the wave tank with the detection window 0.6 m under water. It was operated in real time mode to acquire dynamic particle size distribution and total particle concentrations (2.5–500 μm) every 3 s.

Water samples were collected for observation of dispersed oil droplets under transmitted light and ultraviolet epi-fluorescence illumination using a Leitz Orthoplan microscope equipped with a computer-controlled motorized stage (Lee *et al.*, 1985). Photomicrographs of the naturally- and chemically-dispersed oil droplets and OMAs were recorded at magnifications of 160 × or 400 × and quantified using image analysis software (Image-Pro5.0). Enumeration results were reported as number of particles per unit volume of water samples.

## 3.  Results and Discussion

### 3.1. OIL-MINERAL AGGREGATION IN THE WAVE TANK

Distribution of crude oil in the wave tank water column as a result of dispersion was quantified by three indicators: the UVF analysis of emission spectra that are representative of aromatic components of the dispersed and/or aggregated oil; the GC-FID analysis of total petroleum hydrocarbon (TPH); and the LISST quantification of total particle concentration.

All the UVF spectra were plotted after correction for natural fluorescence and light scattering of blanks made up of seawater used in each wave tank treatment. Figure 2 shows the emission spectra of the crude oil dispersed at different depths at 1.5 m downstream from oil addition after 1 min of breaking waves. The treatment effects on the vertical dispersion of oil at the 1.5 m downstream location were apparent: natural dispersion of oil has extremely limited effectiveness (Figure 2a); the action of dispersant (Figure 2b) or mineral fines (Figure 2c) on their own disperses more oil; the action of mineral fines and dispersant in combination disperses more oil into the bulk aqueous phase (Figure 2d). Samples taken from 4 m downstream from the oil addition site did not exhibit strong UVF signals over all depths after 1 min. After 10 min, emission spectra from samples taken at different depths and horizontal locations became similar for the same treatment, indicating that breaking waves resulted in a complete mixing downstream flow. Hence, the fluorescence intensity was presented as an average value from samples taken at the four different depths and two horizontal locations (eight readings) for each treatment at each time point.

The spectra from the wave tank samples are similar to those obtained from results of previous flask experiments, where the aggregation of oil with mineral fines produces distinct shifts in UVF spectra compared to oil dispersed in seawater or hexane (Kepkay et al., 2002). The UVF spectra emission peaks at various excitation ranges are correlated to the different dissolved/dispersed fractions of crude oil (ASTM, 2002); most single ring aromatics, such as BTEX, phenol, and cresol, produce strong emission peaks between 290 and 320 nm; two-ring aromatics, such as naphthalene and phenanthrene, generate peaks at 325–370 nm; multi-ring compounds, such as resins, form another major broad peak from 400 to 470 nm. Since most of the single ring aromatics have been weathered, and solubility of such compounds high, the effects of dispersants and mineral fines on their partitioning to bulk water phase are probably insignificant.

Figure 3 shows the dispersal of oil with an average fluorescence at 340–370 nm (Figure 3a) and an average fluorescence at 400–480 nm

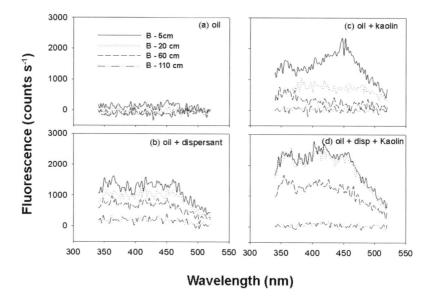

*Figure 2.* Emission spectra of Scotia Light condensate dispersed at 4 depths (5, 20, 60, and 110 cm) 1.5 m downstream 1 min after addition into seawater in combination with low concentrations of kaolin particles and/or oil dispersant under breaking waves.

(Figure 3b) in the wave tank under different treatments over time. The average fluorescence intensity of oil fraction that fluoresces at 340–370 nm followed the general order: oil < oil + kaolin < oil + dispersant < oil + dispersant + kaolin (Figure 3a). In the end, the chemically dispersed oil and the mineral-fines dispersed oil exhibited fluorescence 3.5 and 2 times higher the naturally dispersed oil, and the combination of dispersant and kaolin increased fluorescence by 5.6-fold relative to the natural oil. The dispersal of oil fraction with fluorescence at 400–480 nm generally followed the same pattern; except that kaolin on its own did not increase the fluorescence in the end, whereas dispersants or kaolin plus dispersants increased the fluorescence by three times. These results suggest that both chemical dispersants and mineral fines stimulated the dispersion of two-ring and multiple-ring aromatic fractions of oil into bulk aqueous phase. The effect of dispersants is caused by the reduction of water-oil interfacial tension in the presence of dispersants. The interaction of kaolin and oil may have reduced the negative buoyancy of oil-droplets and therefore transfer more oil into bulk aqueous phase. The combination of dispersants and mineral fines appears to have a synergetic effect on transferring oil to dissolved/ dispersed phases.

The distribution of oil in wave tank as a result of dispersal by dispersants and mineral fines were further quantified with GC-FID analysis of TPH from samples taken from four depths and two locations. Like UVF results, TPH concentrations from both the dissolved and sequestered phases showed no temporal and spatial difference among these samples taken at the same time point after 10 min due to the mixing effect in the wave tank. Hence the average concentrations of the TPH in the dissolved seawater, in the sequestered solid phase, and the total suspended TPH from two horizontal locations and four depths (eight readings) was calculated, and the results are presented in Figure 4.

The presence of chemical dispersants significantly enhanced the concentration of dissolved oil in wave tank (Figure 4a); the presence of mineral fines appears to have reduced the amount of the soluble oil in seawater, and the combination of dispersants and mineral fines significantly reduced the dissolved oil concentration in bulk aqueous phase. Conversely, the presence of mineral fines and chemical dispersants both stimulated the sequestration of total petroleum hydrocarbon in solid phase (Figure 4b). This is not surprising because the mineral fines are well known to absorb oil at their surface due to their hydrophobic surface properties. Likewise, because dispersants have both hydrophobic and hydrophilic ends, they may also function to reduce the surface tension between oil and water, and then stimulate the interaction between oil and mineral fines. Due to the relatively significant amount of suspended particles in bulk aqueous phase, the net effect of mineral fines and dispersants on distribution of oil in bulk water will significantly increase the total suspended oil concentration in bulk water (Figure 4c).

The presence of chemical dispersants significantly enhanced the concentration of dissolved oil in wave tank (Figure 4a); the presence of mineral fines appears to have reduced the amount of the soluble oil in seawater, and the combination of dispersants and mineral fines significantly reduced the dissolved oil concentration in bulk aqueous

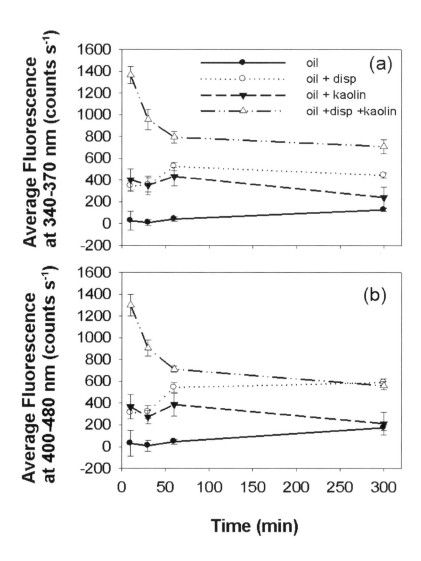

*Figure 3.* UVF measurement of (a) dispersed oil fraction with average fluorescence at 340–370 nm, and (b) oil fraction with average fluorescence at 400–480 nm over time. Results shown are the tank average value for each time point and error bars represent one standard deviation of eight samples.

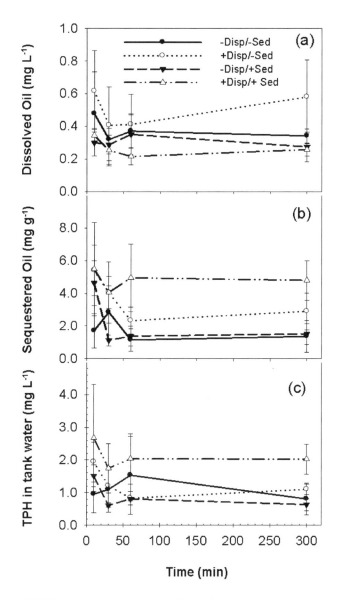

*Figure 4.* GC-FID measurement of: (a) dissolved oil (micrograms per liter of seawater), (b) aggregated oil (micrograms per gram of particles), and (c) total suspended oil in dissolved and sequestered phases over time. Results shown are the tank average value for each time point and error bars represent one standard deviation of eight samples.

phase. Conversely, the presence of mineral fines and chemical dispersants both stimulated the sequestration of total petroleum hydrocarbon in solid phase (Figure 4b). This is not surprising because the mineral fines are well known to absorb oil at their surface due to their hydrophobic surface properties. Likewise, because dispersants have both hydrophobic and hydrophilic ends, they may also function to reduce the surface tension between oil and water, and then stimulate the interaction between oil and mineral fines. Due to the relatively significant amount of suspended particles in bulk aqueous phase, the net effect of mineral fines and dispersants on distribution of oil in bulk water will significantly increase the total suspended oil concentration in bulk water (Figure 4c).

The effects of dispersants and mineral fines on oil dispersion or distribution in wave tank are also illustrated in Figure 5, where the partitioning ratios of oil between filter-retained solid phase and liquid phase were plotted for different treatments. Under natural dispersion, the solid to liquid ratio was probably due to the ratio of insoluble components of the TPH to the soluble fraction of the TPH. The presence of either mineral fines or dispersants alone significantly increased the partitioning ratio, suggesting that either part of the insoluble TPH has been dispersed by the presence of chemical dispersants, or part of the TPH was absorbed by mineral fines and hence the gravity density was increased to overcome the buoyancy. The presence of both dispersants and mineral fines dramatically increased the partitioning ratio of TPH between solid phase and dissolved phase, indicating that the combination of the two has synergetic effects on stimulating the intrusion of oil into bulk aqueous phase.

Figure 6 presents time-series total particle concentration measured by LISST. In the absence of mineral fines, the total dispersed oil first increased to peak at 10 min, and then decreased gradually to a relatively consistent level. In the presence of mineral fines, however, the peak concentrations were detected at the start and gradually decreased with time. The total particle concentrations in the presence of mineral fines were always twice higher than in their absence, with the final total average concentrations being 10 and 4 $\mu l \cdot L^{-1}$, respectively. Chemical dispersants have not affected total suspended particle concentrations, presumably due to the rapid mixing of high energy dissipation rate breaking waves.

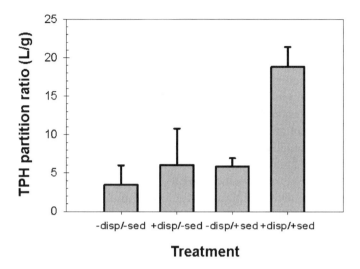

*Figure 5.* Treatment effects on the partitioning ratio of oil between dissolved phase and solid phase.

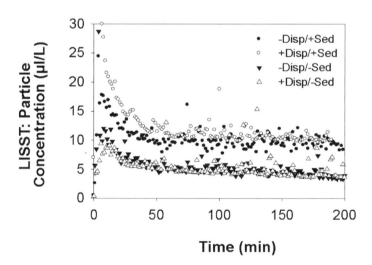

*Figure 6.* LISST measurement of total particle concentration of dispersed oil and OMAs.

## 3.2. PARTICLE SIZE DISTRIBUTION

The dynamic change of particle size distribution was detected using the
LISST-100X, an *in-situ* laser emission and detection particle size ana-
lyzer. Figure 7 shows the continuous time-series particle size distribution expressed as mass mean diameter (MMD) of the particles. The
presence of dispersant had a significant effect on the particle size distribution: the natural (physical) dispersion of oil in the absence of
either dispersant or mineral fines first produced large droplets (MMD ~
150 μm) and gradually reduced (MMD ~ 40 μm) after continuous dispersion under breaking waves for 200 min. The presence of mineral
fines reduced the initial particle size by almost 100% (MMD ~ 70 μm),
but the formed OMAs appear to become stabilized and did not break
up further. The presence of dispersant dramatically reduced the sizes
of both dispersed oil droplets (treatment + disp/–kaolin) and OMAs
(treatment + disp/+kaolin) sizes to MMD of 15–25 μm.

Naturally dispersed oil has the largest oil droplet size initially, with
more than 50% of the dispersed oil volume present as droplets larger
than 200 (mass medium diameter, or $d_{50\%}$), and a geometric standard
deviation (GSD) of 1.90. Further dispersion of oil by the breaking

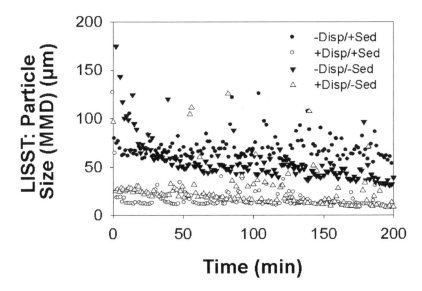

*Figure 7.* LISST measurement of mass mean particle diameter of dispersed oil and
OMAs.

wave turbulence gradually reduced the oil droplet size over time. The mass medium diameter of the droplet size was almost an order of magnitude lower than the initial size.

The presence of chemical dispersants dramatically decreased the initial droplet size, with $d_{50\%}$ being only 22 μm, about 10% of the value in the absence of dispersants. Further dispersion of oil in the presence of chemical dispersants in the wave tank generated large amount of small particles after 5 h (e.g., 84% of droplets are less than 11 μm and 50% are less than 3 μm). The presence of mineral fines also signify-cantly decreased the initial particle size compared to natural dispersion of oil; however, the formed oil-mineral aggregates appear to be more stable than their counterparts of the dispersed oil droplets. The stability of OMA structure that has been observed in this study is consistent with previous reports (Khelifa et al., 2002; Muschenheim and Lee, 2002; Stoffyn-Egli and Lee, 2002). The presence of both chemical dispersants and mineral fines produced the smallest initial particle size (e.g., mass medium diameter was 10.5 μm, only 5% of the naturally dispersed oil). Like the physically dispersed OMAs, once they were generated, these particles appeared to be relatively stable in the wave tank throughout the experiment under breaking-wave turbulence.

The physically and chemically dispersed OMAs were observed under microscope installed with transmitted light and epi-fluorescent illumination. Consistent with the LISST results, the sizes of the OMAs formed by chemically dispersed oil are much smaller than the physi-cally dispersed. In morphology, the chemically dispersed OMAs appe-ared to have predominantly fractal structures, whereas the physically dispersed OMAs were largely spherical. Enumeration of the dispersed oil and OMA particles under microscope indicates that either chemical dispersants or mineral fines on their own significantly increased the particle number per unit volume.

## 4. Conclusions

The results presented here from a wave tank study demonstrated that dispersants and mineral fines have significant impact on the formation and transport of oil-droplets and OMAs, such as intrusion of oil in bulk aqueous phase and particle size distribution. The formation and

transport of oil droplets and OMAs in seawater as a result of interactions between crude oil, chemical dispersant and mineral fines produced distinct effects under breaking waves that were regulated by the
treatment used. The effects were as follows: the interaction of chemical
dispersant with oil and mineral fines increased the dissolved and aggregated oil concentration in the bulk aqueous phase and reduced the size
of OMAs to a mass mean diameter of 15–25 μm. OMAs formed in the
absence of dispersant also exhibited a stable distribution in suspension
but their particle sizes were larger than naturally dispersed oil droplets
after prolonged (5 h) mixing under the breaking-wave hydrodynamic
regime. Natural dispersion of crude oil is of limited effectiveness;
however, prolonged action of the breaking-waves improved the natural
dispersion by progressively reducing the dispersed oil-droplet size.
Chemical dispersion of the same oil promoted the entrainment of the
oil into the bulk aqueous phase and decreased the size of dispersed oil
droplets over time. The impacts of the interaction among chemical dispersant, mineral fines and crude oil on the number of dispersed oil or
OMAs and their stability in suspension were confirmed by observation
via microscope.

## 5.  Acknowledgements

This research was supported by the Panel of Energy Research and
Development (PERD) Canada, U.S. EPA (research contract No. 68-C-
00-159), and NOAA/UNH Coastal Response Research Center (Grant
Number: NA04NOS4190063 UNH Agreement No.: 06-085). Essential
technical and logistical support to this research program was provided
by Susan Cobanli, Jennifer Dixon, Xiaowei Ma, Peter Thamer, Matt
Arsenault, William Yeung, and Jay Bugden.

## References

Anderson, C.M. and LaBelle, R.P., 2000, Update of comparative occurrence rates for
    offshore oil spills. *Spill Science and Technology Bulletin*, **6**(5–6), 303–321.
ASTM, 2002, *American Society for Testing and Materials. Standard test method for
    waterborne petroleum oils by fluorescence analysis*. ASTM D3650-90, West
    Conshohocken, PA.

Boufadel, M.C., Wickley-Olsen, E., Kaku, V.J., King, T., Li, Z., Lee, K., and Venosa, A.D., in preparation, Regular and breaking waves in wave tank for dispersion effectiveness testing: 1 characterization of hydrodynamics.

Bragg, J.R. and Owen, E.H., 1995, Shoreline cleansing by interactions between oil and fine mineral particles. *1995 International Oil Spill Conference*, 219–227.

Bragg, J.R. and Yang, S.H., 1995, *Clay-oil flocculation and its effects on the rate of natural cleansing in Prince William Sound following the Exxon Valdez oil spill. Exxon Valdez Oil Spill – Fate and Effects in Alaskan Waters*, P.G. Wells, J.N. Butler, and J.S. Hughes, eds., American Society for Testing and Materials, Philadelphia, PA, 178–214.

Cloutier, D., Amos, C.L., Hill, P.R., and Lee, K., 2002, Oil erosion in an annular flume by seawater of varying turbidities: A critical bed shear stress approach. *Spill Science and Technology Bulletin,* **8**(1), 83–93.

Delvigne, G.A.L. and Sweeney, C.E., 1988, Natural dispersion of oil. *Oil and Chemical Pollution*, **4**(4), 281–310.

Delvigne, G.A.L., Van del Stel, J.A., and Sweeney, C.E., 1987, *Measurements of vertical turbulent dispersion and diffusion of oil droplets and oil particles.* MMS 87–111, US Department of the Interior, Minerals Management Service, Anchorage, Alaska.

EPA, 2003, *US Environmental Protection Agency. Test Methods* (2003). EPA Methods 3510C, 3540C and 8100.

Gearing, J.N. and Gearing, P.J., 1983, Suspended load and solubility effect on sedimentation of petroleum hydrocarbons in controlled estuarine ecosystems. *Canadian Journal of Fisheries and Aquatic Science*, **40**, 54–62.

Guyomarch, J., Le Floch, S., and Merlin, F.X., 2002, Effect of suspended mineral load, water salinity and oil type on the size of oil-mineral aggregates in the presence of chemical dispersant. *Spill Science and Technology Bulletin*, **8**(1), 95–100.

Guyomarch, J., Merlin, F., and Bernanose, P., 1999, Oil interaction with mineral fines and chemical dispersion: Behaviour of the dispersed oil in coastal or estuarine conditions. Environment Canada's 22nd Arctic and Marine Oilspill (AMOP) Technical Seminar, Calgary, Alberta, Canada, pp. 137–149.

Kepkay, P.E., Bugden, J.B.C., Lee, K., and Stoffyn-Egli, P., 2002, Application of ultraviolet fluorescence spectroscopy to monitor oil-mineral aggregate formation. *Spill Science and Technology Bulletin*, **8**(1), 101–108.

Khelifa, A., Stoffyn-Egli, P., Hill, P.S., and Lee, K. 2002. Characteristics of oil droplets stabilized by mineral particles: Effects of oil type and temperature. *Spill Science and Technology Bulletin,* **8**(1), 19–30.

Le Floch, S., Guyomarch, J., Merlin, F.X., Stoffyn-Egli, P., Dixon, J., and Lee, K., 2002, The influence of salinity on oil-mineral aggregate formation. *Spill Science and Technology Bulletin*, **8**(1), 65–71.

Lee, K., 2002, Oil-particle interactions in aquatic environments: Influence on the transport, fate, effect and remediation of oil spills. *Spill Science and Technology Bulletin*, **8**(1), 3–8.

Lee, K., Wong, C.S., Cretney, W.J., Whitney, F.A., Parson, T.R., Lalli, C.M., and Wu, J., 1985, Microbial response to crude oil and Corexit 9527: SEAFLUXES enclosure study. *Microbial Ecology*, **11**, 337–351.

Lee, K., Weise, A.M., and St-Pierre, S., 1996, Enhanced Oil Biodegradation with Mineral Fine Interaction. *Spill Science and Technology Bulletin*, **3**(4), 263–267.

Lee, K., Lunel, T., Wood, P., Swannel, R., and Stoffyn-Egli, P., 1997, Shoreline cleanup by acceleration of clay-oil flocculation process. *The 1997 International Oil Spill Conference*, 235–240.

Lee, K. and Stoffyn-Egli, P., 2001, Characterization of oil-mineral aggregates, in: *Proceedings of the 2001 International Oil Spill Conference*. American Petroleum Institute, Washington, DC, pp. 991–996.

Lee, K., Stoffyn-Egli, P., Tremblay, G.H., Owens, E.H., Sergy, G.A., Guenette, C.C., and Prince, R.C., 2003, Oil-mineral aggregate formation on oiled beaches: Natural attenuation and sediment relocation. *Spill Science and Technology Bulletin*, **8**(3), 285–296.

Li, M. and Garrett, C., 1998, The relationship between oil droplet size and upper ocean turbulence. *Marine Pollution Bulletin*, **36**, 961–970.

Lunel, T., Swannell, R., and Rusin, J., 1997, Monitoring the effectiveness of response operations during the Sea Empress incident: a key component of the successful. *Oceanographic Literature Review*, **44**(12), 1570–1570.

MacKay, D. and Hussain, K., 1982, An exploratory study of sedimentation of naturally and chemically dispersed oil. *Environment Canada Report*, EE-35, 24 p, Ottawa, Ontario, Canada.

Muschenheim, D.K. and Lee, K., 2002, Removal of oil from the sea surface through particulate interactions: Review and prospectus. *Spill Science and Technology Bulletin*, **8**(1), 9–18.

NRC, 2003, *National Research Council: Oil in the Sea III: Inputs, Fates and Effects.*, National Academies Press, Washington, DC.

Omotoso, O.E., Munoz, V.A., and Mikula, R.J., 2002, Mechanisms of crude oil-mineral interactions. *Spill Science and Technology Bulletin*, **8**(1), 45–54.

Owens, E.H. and Lee, K., 2003, Interaction of oil and mineral fines on shorelines: review and assessment. *Marine Pollution Bulletin*, **47**(9–12), 397–405.

Owens, E.H., Davis Jr., R.A., Michel, J., and Stritzke, K., 1995, Beach cleaning and the role of technical support in the 1993 Tampa Bay spill. *The 1995 International Oil Spill Conference*, 627–634.

Owens, E.H., Sergy, G.A., Guenette, C.C., Prince, R.C., and Lee, K., 2003, The Reduction of Stranded Oil by In Situ Shoreline Treatment Options. *Spill Science and Technology Bulletin*, **8**(3), 257–272.

Page, C.A., Bonner, J.S., Sumner, P.L., McDonald, T.J., Autenrieth, R.L., and Fuller, C.B., 2000, Behavior of a chemically-dispersed oil and a whole oil on a near-shore environment. *Water Research*, **34**(9), 2507–2516.

Shaw, J.M., 2003, A Microscopic View of Oil Slick Break-up and Emulsion Formation in Breaking Waves. *Spill Science and Technology Bulletin*, **8**(5–6), 491–501.

Stoffyn-Egli, P. and Lee, K., 2002, Formation and characterization of oil-mineral aggregates. *Spill Science and Technology Bulletin*, **8**(1), 31–44.

Tkalich, P. and Chan, E.S., 2002, Vertical mixing of oil droplets by breaking waves. *Marine Pollution Bulletin*, **44**(11), 1219–1229.

Venosa, A.D., Kaku, V.J., Boufadel, M.C., and Lee, K., 2005, Measuring energy dissipation rates in a wave tank. In: *Proceedings of 2005 International Oil Spill Conference*, Miami, FL. American Petroleum Institute, Washington, DC.

# PART 3. BIOLOGICAL EFFECTS AND MONITORING HABITAT RECOVERY

# OPTIMAL NUTRIENT APPLICATION STRATEGIES FOR THE BIOREMEDIATION OF TIDALLY-INFLUENCED BEACHES

M.C. BOUFADEL[†], H. LI & C.-H. ZHAO
*Department of Civil and Environmental Engineering,*
*Temple University, Philadelphia, PA 19122, USA*

A.D. VENOSA
*National Risk Management Research Lab, U.S. EPA,*
*Cincinnati, OH 45268, USA*

**Abstract.** Bioremediation of oil spills on tidally-influenced beaches commonly involves the addition of a nutrient solution to the contaminated region of the beach to stimulate the growth of indigenous oil-degrading bacteria. Maximizing the residence time of nutrients in the bioremediation zone (the top portion of the beach) is a main goal for successful bioremediation. Our previous investigations revealed that the predominant flow during rising tides is downward into the beach. In other words, the tide fills the beach from above. Our studies also showed that the predominant flow during falling tides is seaward. This means that the rising tide "flushes" the nutrients from the nutrient zone, and that the falling tide carries them to sea. It became obvious to us that the best application strategy for dissolved nutrients should occur during falling tide and not rising tides. In this work, we applied the nutrients in dissolved form at the high tide line during a falling tide at a high flow that caused the runoff of the applied solution seaward covering therefore the entirety of the intertidal zone by the time the tide reaches the low tide line. Simulation of the results in the MARUN model (a finite element model for beach hydraulics and hydro-dynamics) provided the needed flow rate. For example, for a beach slope of 10%, a tidal range of 2 m, a hydraulic conductivity of $10^{-3}$m/s

---

[†] To whom correspondence should be addressed. E-mail: boufadel@temple.edu.

W. F. Davidson, K. Lee and A. Cogswell (eds.), *Oil Spill Response: A Global Perspective.*   181
© Springer Science + Business Media B.V. 2008

(sand), and a diurnal tidal period, the applied flow rate of the nutrients' solution should be 0.3 L/s resulting in a volume of solution of approximately 3 m $^3$. The results are generalized to other beaches using a novel dimensionless formulation.

**Keywords:** bioremediation, bacteria, tide, oil spill response

# BENTHOS COMMUNITY MONITORING OF A DUMPING AREA DURING LIQUID NATURAL GAS PLANT CONSTRUCTION

A.D. SAMATOV[†] & V.S. LABAY
*Sakhalin Fisheries and Oceanography Research Institute, P.O. Box 693020, Russia, Yuzhno-Sakhalinsk, Komsomolskaya str. 196*

**Abstract.** According to the technical-economic substantiation (TES) of the Project "Sakhalin-II – Phase 2", 2 marine exploration platforms are being constructed, one of them in the Lunskoye gas field. They will be connected by pipelines to the oil terminal in Aniva Bay (south coast of Sakhalin island) for year round exploration. Another structure, a liquefied natural gas (LNG) plant, was also built in 2003 near stm. Prigorodnoye on the Aniva Bay shore. The Contractor for this work is CTSD Ltd. The LNG terminal and a material operations facility (MOF) comprise the LNG plant infrastructure. A dredging operation was needed in the area of the terminals prior to construction to provide ships safe approach and mooring. The dumping of dredged material was approved for Aniva Bay outside of a 12 mile zone where water depth was between 60–65 m. The Project realization includes a complex environmental monitoring program. The primary research objective was to monitor marine biota and the surrounding environment of the dredging and dumping areas in Aniva Bay. One of the main monitoring tasks of the Environmental Impact Assessment (EIA) for the project was to specify the forecast estimation of the impact on the benthic community and the process of restoring the original abundance and structure of the affected area. FSUE "Sakhalin Fishery and Oceano-graphy Research Institute" (SFORI or SakhNIRO) was responsible for conducting this research based on the Agreement with CTSD Ltd. The schedule of observations includes the following phases: before dredging, during dumping and upon completion.

**Keywords:** natural gas, monitoring

[†] To whom correspondence should be addressed. E-mail: samatov@sakhniro.ru

W. F. Davidson, K. Lee and A. Cogswell (eds.), *Oil Spill Response: A Global Perspective.*   183
© Springer Science + Business Media B.V. 2008

## 1.  Materials and Methods

Benthos was sampled in August 2003 – before ground dumping; in October and December 2004, May and August 2005 – during construction; and, in August 2006 after work had been completed. Sampling stations were located at 300, 800 and 2,000 m to the north, east, south and west from the central point of construction (coordinates 46° 24.5,0' N latitude and 142°55,0' E longitude). Benthos was sampled from the R/V "Dmitry Peskov" using a Van-Veen grab (0.2 m²). Benthic samples were placed on a table and washed with outboard water through a sieve system containing mesh 4, 2, and 1 mm in diameter. Organisms were removed from the >1 mm sieves and the accumulated sediment/ organism mixture was fixed for further examination. Small macrobenthos were extracted under laboratory conditions when sediment was washed through a sieve system with mesh diameters of 1 and 0.5 mm. Samples were processed via a binocular microscope MBS-10. The composition, abundance and biomass of each species were determined. Organisms were weighed on the electronic scales AND HM200 up to 0.0001 g; biomass was determined per 1 m².

Different structural indices and coefficients were used for an assessment of the state of, and comparison between, benthic communities:

Index of diversity (Shennon):

$$H' = -\sum b_i/B \log_2 (b_i/B), \qquad (1)$$

where: $b_i$ – biomass of each species, $B$ – total biomass.

INDEX OF CENOTIC SIMILARITY (SHOENER):

$$C_{xy} = 100 - 0{,}5\Sigma(|\, p_x - p_y\,|), \qquad (2)$$

where: $p$ – part (%) of this species in total biomass on x and y stations.

SPECIES SIMILARITY COEFFICIENT (SERENSEN):

$$I_{xy} = 2c * 100/(a + b), \qquad (3)$$

where: $c$ – number of common species for $x$ and $y$ stations; $a$ and $b$ – number of species on $x$ and $y$ stations.

COMMUNITY SUCCESSION INDEX (ABC-METHOD):

$$I_{ABC} = \frac{\sum_{1}^{1W} Bc_i - \sum_{1}^{W} Nc_i}{W}, \quad (4)$$

where: $Bc_i$ – cumulative biomass of $i$ species ($Bc_1$ – biomass of dominate species (%); $Bc_2$ – amount biomass of dominate and next for them species, etc.); $Nc_i$ – cumulative abundance of $i$ species; $W$ – number of species.

## 2.   Results

Dredging and dumping operations were carried out from March until mid-May and from mid-September until the end of December 2004–2005 and during this time the environment and biological resource impact was assessed.

In August 2003, prior to the dumping operation, the benthic community was dominated by sipunculids' (*Golfingia margaritacea*) and a great number of polychaetes (*Axiothella catenata, Praxillella* spp., and *Prionospio* sp.). In addition, there was also a relatively high abundance of amphipods and cumacea. Average benthos biomass was 53.7 g/m$^2$ (Table 1).

In December 2004, after the first phase of dumping, the species list and abundance were elevated by the presence of small crustaceans, but mean biomass was half that measured in August 2003 (Table 1). The bivalve *Nuculana pernula pernula* prevailed over the majority of the area in December 2004. Correspondingly, this type of community structure could be considered indicative of the dumping area.

In August 2005 the mean benthos biomass was rather high – 48.6 g/m$^2$ (Table 1); the biomass consisted primarily of sipunculids (71%). Nonetheless, it should be noted that almost the entire sipunculids biomass was accumulated at the station located 2,000 m westward of the

TABLE 1. Comparisons of benthic quantitative indices in the dumping area at different time periods.

| Time periods | Quantitative indices | | | |
|---|---|---|---|---|
| | Length of species list | N (ind./m$^2$) | B (g/m$^2$) | B,g/m$^2$ (within 300 m) |
| August 2003 | 36 | 200 | 53.7 | 53.7 |
| October 2004 | 17 | 103 | 26 | 26 |
| December 2004 | 35 | 205 | 26.3 | 24.5 |
| May 2005 | 15 | 74 | 5.3 | 5.3 |
| August 2005 | 39 | 120 | 48.6 | 6.5 |
| August 2006 | 35 | 68 | 18.2 | 17.5 |

central dumping point (413 g/m$^2$). Benthos community at this station can be characterized as refugial. At the remaining stations, sipunculids' were infrequently present, and the total benthic biomass did not exceed 36 g/m$^2$. The benthos biomass in dumping zone in August (6.5 g/m$^2$), as in June (5.3 g/m$^2$), were very low as compared to the August 2003 sampling period, this may be considered as a result of impact on the bottom community during the second dumping phase.

Interestingly, we noted a rather high biomass of bivalves located 300 m northward of the dumping point (3.5 g/m$^2$), which consisted primarily of the common shallow mollusks *Callista brevisiphonata* and *Turtonia minuta.* Finding these and other shallow water species (e.g., green algae and the isopod *Arcturus crassispinis*) leads one to suspect that there was extracted nearshore ground discharged at the site.

The mean benthos biomass in August 2006 was 18.2 g/m$^2$; the majority of biomass (69%) was comprised of bivalves. Distribution irregularities were observed in the benthic group's biomass. The bivalve *Nuculana pernula pernula* prevailed at stations located nearby the dumping point (300 m to the west, east and north), and also to the north (800 and 2,000 m) (Figure 1). Correspondingly, the post-dumping community is observed in this area.

The maximum polychaete biomass was observed at a distance of 2,000 and 800 m to the west off the dumping point, and 2,000 m to the east north off it (Figure 2); sipunculids were also present at 800 m to the west and east off the Dumping point (Figure 3). Therefore the bottom community at these stations could be characterized as recovering pre-dumping (refugial).

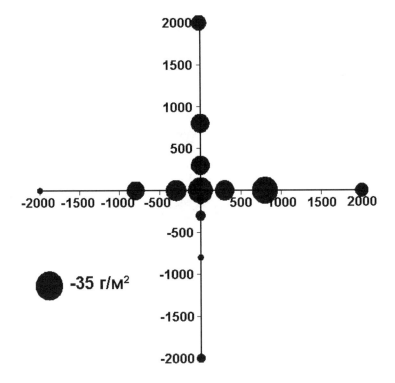

*Figure 1.* Bivalves biomass distribution.

On the whole, within 300 m radius of the dumping point to the west, north and east there were observed rather high biomasses (from 13 to $30g/m^2$); this could be considered as a result of the gradual restoration of bottom biota after dumping. The mean biomass within 300 m radius of the Dumping point was $17.5g/m^2$, that value is higher than the indices of May and August 2005.

Now we compare benthos structure characteristics at the dumping area by the study periods (Table 2). In August 2005 the community structure was close to that of the background phase; this is demon-strated by the Shennon's and ABC-index. The level of cenotic and species similarity (Serensen's coefficient) with the background commu-nity was also high; this proves a gradual recovery of the post-dumping benthic community up to the background one, although the presence of bivalves' is of great significance.

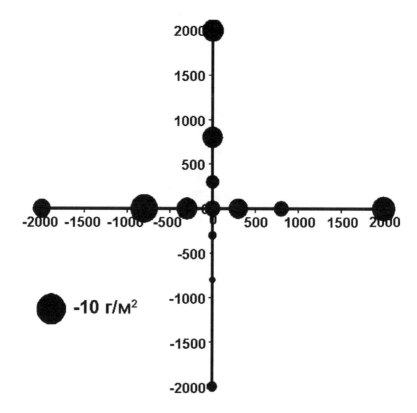

*Figure 2.* Polychaetes biomass distribution.

TABLE 2. Comparison of benthic structural characteristics on the Dumping area by different time periods.

| Time periods | Shennon's index Abundance | Biomass | ABC-index | Cenotic similarity (%) | Serensen's coefficient (%) |
|---|---|---|---|---|---|
| Aug 2003 | 1.341 | 0.793 | 35.6 | 50.1 | 40.5 |
| Oct 2004 | 0.708 | 0.105 | 26.4 | 71.1 | 8.9 |
| Dec 2004 | 0.807 | 0.674 | 17.0 | 47.7 | 36.1 |
| May 2005 | 0.944 | 0.252 | 22.5 | 20.9 | 44.4 |
| Aug 2005 | 1.251 | 0.452 | 30.7 | – | – |

*August 2006 in progress

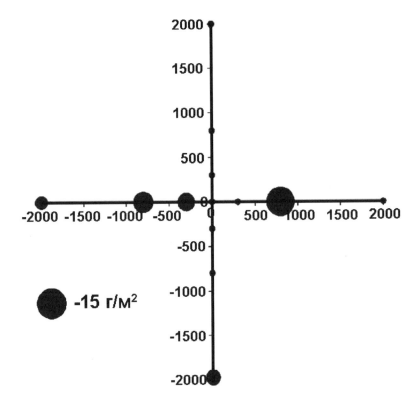

*Figure 3.* Sipunculids biomass distribution.

## 3. Conclusion

The negative dumping impact on bottom community was less significant than expected, and the process of community recovery to the
pre-dumping state is progressing rapidly.

# OIL FINGERPRINTING AND SPILL SOURCE IDENTIFICATION

Z. WANG[†]

*Emergencies Science and Technology Division, Environmental Technology Centre, Environment Canada, 335 River Road, Ottawa, Ontario, K1A 0H3, Canada*

**Abstract**[*]. Oil spills pose a range of environmental risks and causes wide public concern, whether as catastrophic spills or chronic discharges. Therefore, to unambiguously characterize, identify and quantify spill oil hydrocarbons is extremely important for environmental damage assessment; understanding the fate, behaviour and predicting the potential long-term impact of spilled oil on the environment; selecting appropriate spill response measures; and, helping to settle legal liability. The oil fingerprinting and data interpretation techniques briefly discussed in this presentation include the following:

1.  Oil spill historic perspective

2.  Development of oil hydrocarbon fingerprinting techniques

3.  Factors controlling the chemical fingerprints of spilled oil:

    –   Primary controls: crude oil genesis – composition of crude oil

    –   Secondary controls: petroleum refining – chemical composition of refined products

    –   Tertiary controls: effects of weathering

4.  Oil correlation and source identification by PAH fingerprinting analysis:

    –   PAH distribution pattern recognition

    –   Diagnostic ratios of PAHs

    –   PAH isomer and cluster PAH analysis

    –   Characterization of additives for source identification of refined products

---

[†] To whom correspondence should be addressed. E-mail: zhendi.wang@ec.gc.ca
[*] Full presentation available in PDF format on CD insert.

W. F. Davidson, K. Lee and A. Cogswell (eds.), *Oil Spill Response: A Global Perspective.*   191
© Springer Science + Business Media B.V. 2008

5. Distinguishing biogenic and pyrogenic hydrocarbons from petro-
   genic hydrocarbons

6. Fingerprinting biomarkers for oil correlation and source identi-
   fication:
   - Tri- to penta-cyclic biomarker distributions in oils and refined
     products
   - Unique biomarkers
   - Weathering check using biomarkers
   - Diagnostic indices and cross-plots of biomarkers
   - Biodegradation of biomarkers
   - Application of biomarkers for spill source identification, oil
     correlation and differentiation

7. Using sesquiterpane and diamondoids for source identification of
   spilled lighter fuels

8. Oil spill identification by statistical and numerical analysis.


**Keywords:** oil spill response, source identification, fingerprinting

# WHAT COMPOUNDS IN CRUDE OIL CAUSE CHRONIC TOXICITY TO LARVAL FISH?

P. HODSON[†], C. KHAN, G. SARAVANABHAVAN,
L. CLARKE, B. SHAW, K. NABETA, A. HELFERTY
& S. BROWN
*Queen's University, 116 Barrie Street, Kingston,
Ontario, K7L 3N6, Canada*

Z. WANG & B. HOLLEBONE
*Environment Canada, Ottawa, Ontario, K1A 0H3,
Canada*

K. LEE
*Fisheries and Oceans Canada, Dartmouth, Nova Scotia,
B2Y 4A2, Canada*

J. SHORT
*Auke Bay Laboratory, National Oceanic
and Atmospheric Administration, 11305 Glacier
Highway, Juneau, AK 99801, USA*

**Abstract.** Early life stages of fish exhibit dioxin-like toxicity when chronically exposed to crude oil. The effects are termed blue sac disease (BSD), and are characterized by edema, haemorrhaging, developmental defects, and the induction of cytochrome P4501A (CYP1A) enzymes. These effects have been correlated to the concentrations of alkyl-substituted polynuclear aromatic hydrocarbons (alkyl-PAH) in oil, but the range of compounds causing toxicity is unknown. To identify the compounds in oil that cause toxicity, we measured the relative potency of four crude oils and sub-fractions of oil for causing CYP1A induction in juvenile trout and BSD in larval trout and medaka. Identification of compounds will be based on correlations between measured potency and chemical analyses of specific compounds or classes of compounds

---

[†] To whom correspondence should be addressed. E-mail: hodsonp@biology.queensu.ca

W. F. Davidson, K. Lee and A. Cogswell (eds.), *Oil Spill Response: A Global Perspective.* 193
© Springer Science + Business Media B.V. 2008

in each fraction. Four oil fractions created by low temperature vacuum distillation varied widely in their potencies, but only fraction F3, which contained primarily 3–5-ringed alkyl-PAH, waxes, and resins, was highly toxic and induced CYP1A activity. Cold acetone (–80°C) extraction of F3 generated an extract (F3-1) that was rich in CYP1A-inducing compounds and highly toxic to fish; the precipitate (F3-2; mostly wax and resins) had no effects on fish. We further separated F3-1 into 5 sub-fractions by preparative-scale normal-phase HPLC. The late-eluting fractions F3-1-3 and F3-1-4 (3-5-ringed alkyl-PAH), and F3-1-5 (5–7-ringed un-substituted PAH, residual waxes and resins) were potent inducers of CYP1A activity. In contrast, the early-eluting fractions (F3-1-1 and F3-1-2, comprised of residual naphthalene, alkyl-naphthalenes, and dibenzothiophenes) did not induce CYP1A activity. We are currently testing the HPLC fractions, and predict that the only toxic fractions will be F3-1-3 and F3-1-4. Chemical analysis by GC-MS will provide detailed descriptions of fraction constituents and a clearer idea of which compounds might be the toxic constituents. These data will be useful for ecological risk assessments and natural resources damage assessments of complex mixtures of hydrocarbons.

**Keywords:** toxicity, larval fish, crude oil

# POTENTIAL IMPACTS OF AN ORIMULSION SPILL ON MARINE (ATLANTIC HERRING; *Clupea harengus*) AND ESTUARINE (MUMMICHOG; *Fundulus heteroclitus*) FISH SPECIES IN ATLANTIC CANADA

S. COURTENAY[†] & M. BOUDREAU
*Fisheries and Oceans Canada at the Canadian Rivers Institute, Department of Biology, University of New Brunswick, Bag Service #45111, 10 Bailey Drive, Fredericton, New Brunswick, E3B 6E1*

K. LEE
*Center for Offshore Oil and Gas Environmental Research, Fisheries and Oceans Canada, Dartmouth, Nova Scotia, B2Y 4A2, Canada*

P. HODSON & M. SWEEZEY
*Queen's University, 116 Barrie Street, Kingston, Ontario, K7L 3N6, Canada*

**Abstract[*].** The growing potential of emulsified bitumal products for fuel such as orimulsion (an emulsion of 70% bitumen in 30% water) warrants further assessment of their possible environmental impacts associated with spills. In the event of a spill at sea orimulsion may contact animals throughout the water column rather than only at the water surface as expected by conventional heavy fuel oils. In this study we tested orimulsion toxicity during the embryonic development of an estuarine (mummichog; *Fundulus heteroclitus*) and a marine (Atlantic herring; *Clupea harengus*) fish species in duplicate assays for each species. Air injection and varying salinities were included in the herring assays to examine their effects on orimulsion toxicity. Water-accommodated fractions (WAF) of No. 6 fuel oil were also tested in the mummichog assays to compare orimulsion toxicity to that of a heavy fuel

---

[†] To whom correspondence should be addressed.  E-mail: courtenays@dfo-mpo.gc.ca

[*] Full presentation available in PDF format on CD insert.

W. F. Davidson, K. Lee and A. Cogswell (eds.), *Oil Spill Response: A Global Perspective.*    195
© Springer Science + Business Media B.V. 2008

oil. Significant impacts were observed at the lowest tested concentration of 0.001% orimulsion in both species. In the more sensitive of the two species, herring, this concentration produced 100% abnormal larvae. Similar abnormalities were produced in both herring and mummichog, including reduced growth, pericardial edema and spinal deformities. These are also the same types of abnormalities produced by heavy fuel oils and PAHs. The initial and most prominent abnormality was peri-pericardial edema, which was usually accompanied by haemorrhaging at its base in mummichog. Orimulsion-exposed fish also suffered from increased mortality, reduced heart rates, premature hatch and reduced lengths. The toxicity of orimulsion was over 300x greater than #6 fuel oil (WAF). Although effect of salinity on orimulsion toxicity in herring was unclear, air injection greatly reduced toxicity.

**Keywords:** orimulsion, herring, mummichog, toxicity

# EFFECTIVENESS OF DISPERSANTS FOR COASTAL HABITAT PROTECTION AS A FUNCTION OF TYPES OF OIL AND DISPERSANT

Q. LIN[†] & I.A. MENDELSSOHN

*Wetland Biogeochemistry Institute, School of the Coast and Environment, Louisiana State University, Baton Rouge, LA 70803, USA*

**Abstract.** Oil spills in nearshore environments may eventually move into sensitive coastal habitats such as coastal marshes, and could impact marsh organisms. Application of dispersants to spilled oils in nearshore environments before the oil drifts into marshes was simulated, and effectiveness of dispersants' relief of the impact of different oil types on salt marsh plants were investigated. The application of the dispersant JD-2000 significantly relieved the adverse effects of both No. 2 fuel oil and South Louisiana crude oil on the dominant salt marsh plant, *Spartina alterniflora*. Upon contact with plant leaves during the rising tide, the oils sharply reduced the photosynthetic rates of the plant, and increased the percentage of dead tissue. In contrast, the dispersed oils did not significantly affect the marsh plants compared to the no-oil control. However, the effectiveness of the relief of Corexit 9500 on the impact of crude oil was not as great as that of No. 2 fuel oil, although Corexit 9500 also significantly mitigated the impact of crude oil on the salt marsh plant. The current study indicates that dispersants are likely to be most effective in mitigating light fuel oil rather than the viscous crude oil, and have the potential as an alternative countermeasure sure to protect sensitive coastal habitats during nearshore oil spills.

**Keywords:** dispersants, habitat protection, oil spill response, nearshore

---

[†] To whom correspondence should be addressed. E-mail: comlin@lsu.edu

W. F. Davidson, K. Lee and A. Cogswell (eds.), *Oil Spill Response: A Global Perspective.*   197
© Springer Science + Business Media B.V. 2008

# HOW CLEAN IS CLEAN?: DEVELOPMENT OF MONITORING METHODS TO DETERMINE ENVIRONMENTAL AND SOCIALLY ACCEPTABLE END-POINTS OF CLEAN-UP

J. FEJES[†], A. MARTINSSON, AND E. LINDBLOM
*Oil Spill Advisory Service, IVL Swedish Environmental Research Institute Ltd., Business and Development Area Water, Box 21, Stockholm, 060 SE-100 31, Sweden*

**Abstract.** Endpoints for oil spill clean-up have been discussed and evaluated in several projects and Workshops over the last few decades. The results produced have been used to modify and develop more environmentally friendly clean-up methods. Nonetheless, there is still a need for advice and robust monitoring tools necessary for the decision making process when deciding to terminate a clean-up operation. This project has developed three monitoring methods to assess environmental and socially acceptable end-points of clean-up operations for Swedish coastal areas. In general, the methods are based on oils stickiness, thickness and bleeding, and have also been developed for ease of use in the field.

Keywords: environmental monitoring, oil spill response

---

[†]To whom correspondence should be addressed. E-mail: Jonas.fejes@ivl.se

W. F. Davidson, K. Lee and A. Cogswell (eds.), *Oil Spill Response: A Global Perspective.*     199
© Springer Science + Business Media B.V. 2008

# PART 4. MODELLING, FATE AND TRANSPORT

# MICROBIAL AND ABIOTIC REMOVAL OF YDROCARBON COMPONENTS IN OIL TANKER BALLAST WATER PROCESSED BY THE ALYESKA BALLAST WATER TREATMENT FACILITY IN PORT VALDEZ, ALASKA

J.R. PAYNE[†]
*Payne Environmental Consultants, Incorporated, 1991 Village Park Way, Suite 206 B, Encinitas, CA 92024, USA*

W.B. DRISKELL
*6536 20th Avenue NE, Seattle, Washington 98115, USA*

J.F. BRADDOCK & J. BAILEY
*University of Alaska – Fairbanks, P.O. Box 755940, Fairbanks, AK 99775, USA*

J.W. SHORT
*Auke Bay Laboratory, Alaska Fisheries Science Center, 11305 Glacier Highway, Juneau, AK 99801, USA*

L. KA'AIHUE & T.H. KUCKERTZ
*Prince William Sound Regional Citizens' Advisory Council, 3709 Spenard Road, Suite 100, Anchorage, AK 99503, USA*

**Abstract***. The Alyeska Pipeline Service Company (APSC) Ballast Water Treatment Facility (BWTF) at the terminus of the Trans-Alaska Pipeline in Port Valdez, Alaska, treats and discharges an average of 9 million gallons per day of oil-contaminated ballast water offloaded from the oil tankers utilizing the Port. This study quantifies the fractions of benzene, toluene, ethylbenzene, and xylene(s) (BTEX), polycyclic

---

[†] To whom correspondence should be addressed.  E-mail: jrpayne@sbcglobal.net

* Full presentation available in PDF format on CD insert.

W. F. Davidson, K. Lee and A. Cogswell (eds.), *Oil Spill Response: A Global Perspective.*   203
© Springer Science + Business Media B.V. 2008

aromatic hydrocarbons (PAH), and saturated hydrocarbons (SHC) being removed at different stages of treatment inside the terminal and evaluates the relative importance of abiotic (volatilization) versus microbial processes. Evaporation is the dominant removal mechanism for BTEX, lower-molecular-weight SHC, and possibly the naphthalenes in the dissolved air flotation (DAF) cells/weirs and in the Splitter Box distributing DAF effluent to the biological treatment tanks (BTTs). Within the BTTs, microbial degradation of BTEX is very efficient and essentially complete midway through the tanks. During the warmer months, SHC biodegradation within the BTTs is also very rapid, but PAH biodegradation is slower with higher alkylated PAH only being partly degraded before the effluent is discharged into Port Valdez, a sill-constricted, subarctic fjord. Both SHC and PAH biodegradation are limited within the BTTs during colder months. Rate constants for parent and alkylated PAH are presented for subarctic summer and winter conditions. Alkylated PAH homologues that make up the discharged oil signal have been tracked via mussel and sediment samples from the Long-Term Environmental Monitoring Program (LTEMP) that has detected accidental discharges as well as the seasonally-controlled transport of BWTF-sourced dissolved- and particulate/oil-phase fractions throughout the Port.

**Keywords:** microbial, abiotic removal, Alyeska, biodegradation

# OIL SPILL DRIFT AND FATE MODEL

M. SAYED, M. SERRER, AND E. MANSARD[†]
*Canadian Hydraulics Centre/National Research
Council, 1200 Montreal Rd., M-32, Ottawa, Ontario,
K1A 0R6, Canada*

**Abstract.** This paper describes an operational forecasting model of oil spill drift and fate. The model is aimed at predicting the behaviour of oil spills in the St. Lawrence River, and at providing decision support for planning responses and clean-up operations. A software system links the spill model to environmental data, wind forecasts and hydrodynamic models. The present paper focuses on the formulation and algorithms of the oil spill drift and fate model. Processes of mechanical spreading dispersion, weathering, as well as adhesion to shorelines are included. A discrete parcel approach is also employed to track the spill.

**Keywords:** oil spill drift, modelling, dispersion

## 1. Introduction

Knowledge of oil spill spreading, weathering and interaction with shorelines is important for contingency planning and directing response operations. The National Research Council Canadian Hydraulics Centre (NRC-CHC) has collaborated with Environment Canada (EC) in the development of an operational forecasting model for the St. Lawrence River. The model brings together the mechanics of oil slick behaviour, detailed environmental input, and a software environment that provides linking among various models and utilities for decision support.

Many studies have examined oil spill spreading and the weathering processes that change the chemical and physical properties of the oil. Expressions for oil spreading under the forces of gravity, surface tension, inertia and viscous resistance were first developed by Fay (1971). Numerous subsequent studies have examined and verified those formulas.

---

[†] To whom correspondence should be addressed. E-mail: etienne.mansard@nrc-cnrc.gc.ca

W. F. Davidson, K. Lee and A. Cogswell (eds.), *Oil Spill Response: A Global Perspective.*   205
© Springer Science + Business Media B.V. 2008

Horizontal dispersion of an oil slick also takes place under the action of waves, wind and water current. It is difficult to adequately review the literature that addresses those processes here. However, comprehensive reviews and descriptions of oil spill drift and fate models were given by Spaulding (1988) and by Christiensen (1994). Dispersion of oil spills has also been examined by Shen *et al.* (1993).

As an oil spill drifts and spreads, a number of weathering processes takes place. They include evaporation, dissolution, emulsification, bio-degradation and adhesion to shorelines. Evaporation rates for various oil types were measured, for example, by MacKay and Matsugu (1973) and Stiver and MacKay (1984). Other weathering processes, particularly in cold waters, were thoroughly examined by Payne *et al.* (1991). A thesis by Cantin (1992) reviewed available literature and developed a model of oil spill drift and fate. The approach of Cantin (1992) was used, to a large extent, in the present model.

The objective of the work reported here was to employ the available formulation of oil spreading and fate, and to customize the model for the conditions of the St. Lawrence River. An essential part of the work focused on providing accurate input of water current and wind predictions, as well as shoreline information. The software includes interfaces to hydrodynamic models of the St. Lawrence River, wind forecasts of EC, and Geographical Information Systems (GIS) data of shorelines. The output is designed to provide the users with the information required for decision support concerning clean-up operations. More details of the software system can be found in a report by Serrer *et al.* (1996). The following sections of this paper focus on the oil drift and fate model. The governing equations are reviewed, and the numerical approach is described. Examples of test cases are also shown.

## 2.  Governing Equations

The present formulation follows the general approaches of Cantin (1992) and Shen *et al.* (1993). Additionally it includes a new treatment of oil interaction with shorelines and weed beds, integrates input environmental conditions which are specific for the study area, employs efficient search schemes which allows relatively large numbers of parcels to be used, and includes a customized user interface designed for operational use. Serrer *et al.* (1996) outlined the approach used in the present work. This approach is based on simulating the drift and weathering of oil in response to the environmental conditions. Input to the solver includes the results of hydrodynamic models, wind data, shoreline information, spilled oil properties, and initial spill conditions.

The solution of the mass transport equations of the oil is achieved by using a discrete-parcel approach. The oil is divided into a large number of discrete parcels. Each parcel is assigned several time-dependent attributes including mass, thickness, and spatial coordinates. The parcels are first advected by the ambient water current and wind. Next, random fluctuations are applied to the parcels to account for turbulent diffusion. The diffusion coefficient is determined by local shear velocity and water depth. Oil thickness for each parcel is then determined from the mass and distance of neighbouring parcels. Finally, evaporation is calculated for each parcel. As parcels encounter shore-lines or weed beds, deposition may occur depending on the capacity of the shoreline (or weed beds) to retain oil. This retention capacity is modelled by considering two separate values. The first is an absorption probability, that is, the probability that a parcel will be retained at this location. The second value is a maximum thickness of oil allowed at the current location. The discussion below addresses the governing equations used to model the following processes:

- Advection and diffusion
- Mechanical spreading
- Evaporation
- Shoreline and weed bed adhesion

## 2.1. ADVECTION

The advective velocity has two components. The first is due to the mean wind and currents, and the second accounts for local turbulent diffusion.

$$\vec{V}=\vec{V}_m+\vec{V}_t \tag{1}$$

The mean advective velocity $\vec{V}_m$ of each parcel is calculated as:

$$\dot{V}_m = \alpha_w \vec{V}_w + \alpha_c \vec{V}_c \qquad (2)$$

where $\vec{V}_w$ is the wind velocity at 10 m above the water surface, $\vec{V}_c$ is the depth averaged mean water current, $\alpha_w$ is wind drag coefficient (the default value is 0.03), and $\alpha_c$ is the current drag coefficient (the default value is 1.15).

The velocity component $\vec{V}_t$ accounts for turbulent diffusion fluctuations in the drift velocity. Based on the random-walk analysis Fischer $et$ $al.$ (1979)

$$\vec{V}_t = R_n e^{i\theta'} \sqrt{\frac{4(D_e + D_T)}{\Delta t}} \qquad (3)$$

where $\Delta t$ is the time step, $R_n$ is a normally distributed random number of mean value of 0.0 and standard deviation of 1.0, $\theta'$ is a uniformly distributed random angle between 0 and $\pi$, $D_e$ is a dispersion coefficient due to mechanical spreading, and $D_T$ is a diffusion coefficient.

In rivers, the diffusion coefficient is affected by the shear velocity $\bar{U}$ and the depth of flow $h$, Fischer $et$ $al.$ (1979) and Sayre and Chang (1969).

$$D_T = 0.6h\bar{U} \qquad (4)$$

The diffusion coefficient $D_T$ for surface dispersants can be written using Manning's constant, $n_b$ Shen $et$ $al.$ (1993)

$$D_T = 0.4 n_b V_c h^{5/6} \sqrt{g} \qquad (5)$$

where $h$ is water depth, and $g$ is the gravitational acceleration.


## 2.2. MECHANICAL SPREADING

The spreading of surface slicks may correspond to one of three regimes defined by Fay (1971) according to the dominant forces, namely: *gravity-inertia, gravity-viscosity*, and *surface tension-viscosity*. A number of studies have determined a spill's spreading rate as a function of time for each regime. The approach of Cantin (1992) is followed here in order to determine the spreading regime and spreading rates. He considers that oil thickness determines the spreading regime. The spreading rate for each regime is accounted for by including an additional diffusion coefficient to be applied to each parcel. The transition between spreading regimes is considered to depend on oil thickness as follows:

Regime 1 – oil thickness is greater than or equal to $E_{1\min}$
Regime 2 – oil thickness is greater than or equal to $E_{2\min}$
Regime 3 – oil thickness is less than $E_{2\min}$. The oil thickness also has a minimum value which may be taken as 10 μm.

The transition values of oil thickness are given by:

$$E_{1\min} = \sqrt{\gamma_w t_e} \qquad (6)$$

and

$$E_{2\min} = \sqrt{\frac{\sigma_n}{\rho_w(1 - sg_{oil})g}} \qquad (7)$$

where $\gamma_w$ is the kinematic viscosity of water, $t_e$ is the exposure time of the parcel in seconds, $\sigma_n$ is the spreading coefficient or net surface tension, $\rho_w$ is the mass density of water, and $sg_{oil}$ is the specific gravity of the oil. Oil thickness has to be calculated in order to determine the spreading regime.

The local thickness is determined by assuming that the mass in each parcel follows a Gaussian distribution with a standard deviation, $\sigma_p$. Contributions from neighbouring parcels are included in the calculations of the thickness for each parcel. The value of the standard deviation for each parcel, $\sigma_p$ is difficult to directly estimate. The observations of Elliot and Wallace (1989), however, indicate that $\sigma_p$ for a parcel is proportional to the standard deviation of the entire slick, as follows:

$$\sigma_p = 0.3\sigma_{Slick} \qquad (8)$$

The standard deviation of the slick can be estimated from the diffusion coefficient and elapsed time using:

$$\sigma_{Slick} = \sqrt{\sigma^2_{Initial} + 2(D_e + D_T)t_e}  \qquad (9)$$

where $\sigma_{Initial}= 0.0$ for a spill from a point source. For existing slicks where the history of the parcel is unknown, the initial $\sigma_p$ is approximated by the parcel radius.

The local thickness of each parcel $T_p$ is calculated using:

$$(10)$$

$$T_p = \sum_{q=1}^{N} \frac{V_q}{2\pi\sigma_q^2} e^{\left[-\frac{r^2}{2\sigma_q^2}\right]}$$

where $V_q$ is the volume of parcel q, $\sigma_q$ is the standard deviation of the neighbouring parcel q, and $r$ is the distance between centers of parcels.

Having determined the spreading regime, the spreading coefficient $D_e$ is calculated as follows:

Regime 1: *gravity-inertia*

$$(11)$$

$$D_e = \frac{K_{2i}^2}{18}\sqrt{\Delta g V_p}$$

Regime 2: *gravity-viscosity*

$$(12)$$

$$D_e = \frac{K_{2v}K_{2t}}{18}\left(\frac{\Delta g V_p^2 \sigma_n^3}{\rho_w^3 \gamma_w^2}\right)^{1/6}$$

Regime 3: *surface tension-viscosity*

$$(13)$$

$$D_e = \frac{1}{18}\left(\frac{10^5 V_p^{3/4} K_{2t}^4 \sigma_n^2}{\pi\rho_w^2\gamma_w^2}\right)^{1/3}$$

where $K_{2i}$ is a proportionality constant = 1.14, $K_{2v}$ is a proportionality constant = 1.45, $K_{2t}$ is a proportionality constant = 2.30, $V_p$ is the current parcel's volume, $\Delta g = \rho_w(1- sg_{oil})g$, and $\gamma_w$ is the kinematic viscosity of water.

## 2.3. EVAPORATION

Oil evaporation Cantin (1992) is a function of temperature, time, surface area and wind speed. The volume fraction evaporated $\Delta F_v$ is expressed as:

$$\Delta F_v = \left[ \frac{KA_p \Delta t}{V_p} \right] e^{\left( 6.3 - \frac{10.3(T_o^o + G_T F_v)}{T_w^o} \right)} \tag{14}$$

where:

$K = 0.0025 V_w^{0.78}$ , and $V_w$ is wind speed (m/s)

$A_p = $ surface area of the parcel (m$^2$)

$\Delta t = $ time step (s)

$V_p = $ instantaneous parcel volume (m$^3$)

$T_o^o = $ initial boiling temperature from the oil catalogue (°K)

$G_T = $ gradient of the distillation curve, from the oil catalogue

$F_v = $ fraction evaporated

$T_w = $ water temperature (°K).

## 2.4. CHANGES IN OIL PROPERTIES WITH AMBIENT CONDITIONS

The density and dynamic viscosity of the oil in a spill changes in response to the ambient temperature and the evaporated fraction. Those variations are expressed by the following formulas:

$$\rho_p = \rho_{p0} + C_1 F_v - C_2 (T^0) \tag{15}$$

$$\mu_p = \mu_{p0} e^{(C_3 F_v)} e^{\left( \frac{C_4}{T^0} \right)} \tag{16}$$

where $\rho_p$ is the oil density, $\rho_{p0}$ is the initial density of the oil, $\mu_p$ is the dynamic viscosity, $\mu_{p0}$ is the initial dynamic viscosity of the oil, $F_v$ is the evaporated fraction of the oil, $T^0$ is the ambient temperature [°K]. Calibration constants are given the symbols $C_1$, $C_2$, $C_3$, and $C_4$.

## 2.5. SHORELINE AND WEED BED ADHESION

As drifting oil encounters shorelines and weed beds, part of the oil may get deposited (or absorbed). The remaining oil continues to drift. Studies of shoreline interaction with oil (Shen *et al.*, 1993) have examined the oil retention capacity of a number of beach types. That capacity is usually expressed in terms of vulnerability indices which correspond to three categories: full retention of oil, full rejection, and partial retention. The estimates, however, appear to be qualitative. There are no data to differentiate between the capacities of the various beaches which are classified as having partial capacity for oil retention.

The fraction of deposited oil is determined here by assigning an "absorption probability" to each shoreline or weed bed type. Those probabilities are in input data files. A probability of 1 means that every parcel which encounters the shoreline is deposited. Such parcels are not allowed to drift, and remain stationary. Alternatively, a probability of zero corresponds to full rejection of the oil. In that case, all parcels encountering a shoreline are returned back into the stream. Also, parcels passing through weed beds of zero absorption probability continue to drift, unhindered by the weeds. The intermediate case of an absorption probability between zero and 1 correspond to partial deposition of the oil. A uniform random number between zero and 1 is generated for each parcel which encounters the shoreline. If the random number is less than the absorption probability, the parcel is deposited. Otherwise, the parcel continues to drift. Weed beds are treated similarly. The present model allows the possibility of specifying a physically based maximum capacity for different areas inside the model domain. This capacity is defined as a maximum allowed thick-ness of deposited oil.

An alternative approach was used by Shen *et al.* (1993) based on using a half-life value to describe the ability of the shoreline to retain oil. In that case all oil parcels are initially deposited upon encountering a shoreline. The volume fraction, which is re-introduced into the stream, is calculated using an exponential function which depends on the half-life of oil retention. Initial simulations using that approach, however, revealed that estimates of rejected oil would depend on the lengths of the arbitrarily chosen shoreline segments. Therefore, the "absorption probability" approach was employed, which avoided that problem.

## 3. Implementation

The computer code, named Particle Oil Spill Model (POSM), is implemented as a Langrangian discrete parcel solution. Each parcel in the simulation represents a small quantity of oil which is independently transported and weathered by environmental forces. Local parcel concentration is used to determine each parcel's local thickness and surface area. The local thickness is required to determine the dispersion coefficient and the surface area is used to calculate the evaporation. Local thickness is also used in determining weed and shoreline adhesion. Each parcel has a set of time dependent attributes including spatial coordinates and volume, which are computed at each time step.

One of the design objectives for POSM was to make it easy to use hydrodynamic data from various sources. Since there are a number of grid schemes by which hydrodynamics are modelled, it was decided that a triangular mesh was the most flexible way of handling spatial data. Both rectangular and curvilinear grid solutions can be easily mapped to a triangular mesh and interpolation of node data such as currents or wind to a point inside a triangle is straightforward.

The spatial reference frame for POSM is therefore a triangular mesh in UTM coordinates. Each node in the mesh can be assigned static or temporally varying properties such as current, water depth, oil absorption probabilities, etc. These properties are spatially and temporally interpolated at the location of each parcel during the course of the simulation.

To allow the user to define a spatially varying wind field, POSM accepts a second triangular mesh in the same UTM coordinate space as described above. This wind grid is typically much coarser than the hydrodynamic grid. Again each node is assigned static or temporally varying wind data. This data is spatially interpolated to each parcel's location. Temporal variations in wind are optionally interpolated allowing the wind data to be supplied at a much larger interval than the simulation time step.

Finally, in order to evaluate oil spill clean up strategies, a simple oil skimmer has been implemented in POSM. This device is defined as a circular area in the simulation domain which retains all parcels encountered. This skimmer has an infinite holding capacity and an infinite input flux capacity. When this device is enabled, POSM generates a separate time-series file containing the volume of oil retained by the skimmer over time.

## 4. Examples

Examples of simulation runs are shown here. They consider idealized cases in order to help with model verification. We note that verification of the model performance is undertaken by EC. That verification cannot be adequately addressed within the scope of the present paper.

### 4.1. AXISYMMETRIC SPILL

This case represents an instantaneous spill in quiescent water. Both wind and water current have zero values. Each parcel is assigned a volume of 1 L, and a total of 10,060 parcels are instantaneously released (i.e., the volume of the spill is 10.06 m$^3$). Properties of Norman Wells Crude were used in the simulation. The spill is assumed to take place over water of 5 m depth and 10$^\circ$ C. The time step is 120 s, and the total duration of the test is 21 h.

The results are illustrated in Figures 1a, b and c by showing positions of the parcels after 3, 12, and 21 h from the release of the spill. Colours of the parcels show the thickness distribution for the spill. Figure 2 shows the evolution of thickness profile along a cross section of the spill. A cursory examination of the resulting extent and thickness values indicates that the results are within the expected range (e.g., compared to the results of Cantin (1992)).

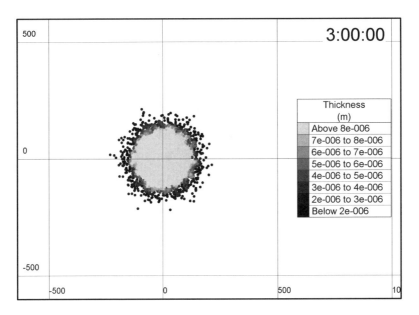

*Figure 1a.* Parcel positions and thickness after 3 h.

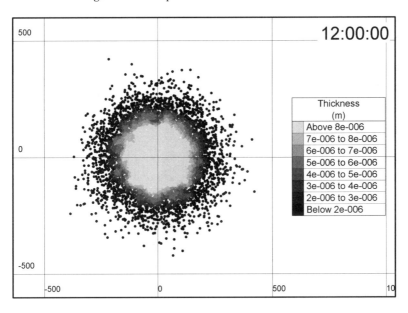

*Figure 1b.* Parcel positions and thickness after 12 h.

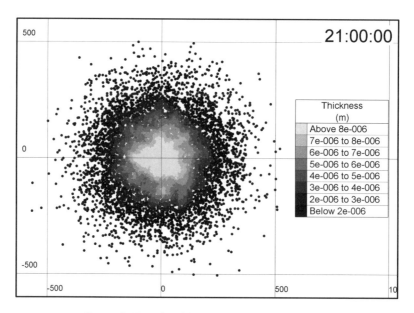

*Figure 1c.* Parcel positions and thickness after 21 h.

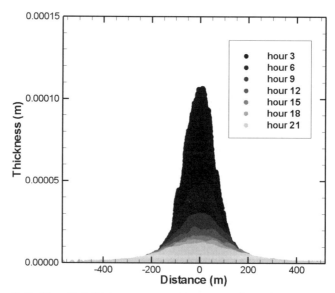

*Figure 2.* Profiles of oil thickness at several increments for an instantaneous axisym-metric spill.

## 4.2.  CONTINUOUS SPILL IN A CHANNEL

This case considers a continuous spill at a rate of 1 L/s in a wide channel of uniform 5 m depth. A water current of 0.15 m/s is assumed to be uniform and steady. Again, properties of Norman Wells Crude were used in the simulation. Water temperature is $10^\circ$ C, and wind velocity is zero. The time step is 60 s.

The resulting positions of the oil parcels are shown in Figures 3a, b and c after 30 min, 2 h, and 3 h. The colours of the parcels give the thickness of the oil. The extent of the spill and dispersion are in general agreement with expected values (e.g., Sayre *et al.,* 1969; Fischer *et al.,* 1979). Quantitative comparisons with measurements of oil spill spreading and other models are conducted by EC, and are beyond the scope of this paper.

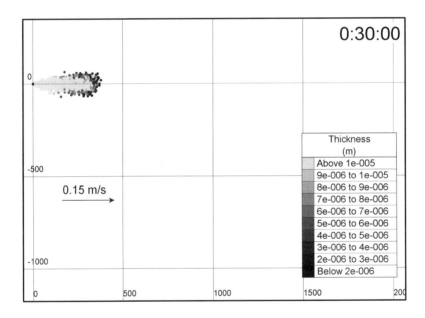

*Figure 3a.* Oil parcel positions and thicknesses after 30 min.

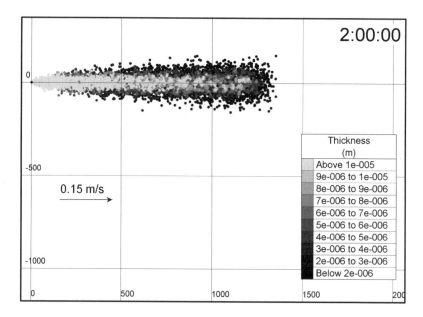

*Figure 3b.* Oil parcel positions and thicknesses after 2 h.

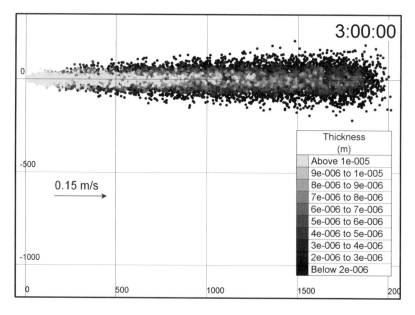

*Figure 3c.* Oil parcel positions and thicknesses after 3 h.

## 5. Conclusions

The present model was developed to provide operational forecasts of oil spills drift and fate in the St. Lawrence River. The model also provides decision support for contingency planning and for directing clean-up operations. The governing equations and algorithms were reviewed in the preceding sections of the paper. Available formulas for mechanical spreading, dispersion, evaporation and adhesion to shorelines were incorporated in the model. A discrete parcel approach was employed to simulate the drift of the oil spill.

The software system was designed to link the oil spill model to hydrodynamics models of the St. Lawrence River, wind forecasts, and GIS data on shorelines. The user interface was also designed to provide the "utility tools" needed for directing response operations.

Validation of the model has been carried out by Environment Canada (EC). The results of such tests and comparisons to available measurements are not included in this paper, but can be directly obtained from EC.

## 6. Acknowledgements

J.F. Cantin provided valuable discussions and suggestions that were instrumental in the development of the model. The support of Environment Canada (EC) is gratefully acknowledged.

## References

Cantin, J.F., 1992, A New Numeric Method for Predicting the Propagation of Oil Spills in Aquatic Environment, Masters thesis, INRS-Eau, Quebec.

Christiensen, F.T., 1994, Management of Oil Spill Risks in Arctic Waters, J. Marine Environ. Eng., **1**, 131–159.

Elliot, A.J. and Wallace, D.C., 1989, Dispersion of Surface Plumes in the Southern North Sea, Dt. Hydrogr. Z., **42**,16.

Fay, J.A., 1971, Physical Processes in the Spread of Oil on a Water Surface, Proc. of the joint Conference on Prevention and Control of Oil Spills, Washington, DC, American Petroleum Institute, pp. 463–467.

Fischer, H.B., List, E.J., Koh, R.C.Y., Imberger, J., and Brooks, N.H., 1979, Mixing in Inland and Coastal Waters, New York: Academic Press.

Mackay, D. and Matsugu, R.S., 1973, Evaporation Rates of Liquid Hydrocarbon Spills On Land and Water, Canadian Journal of Chemical Engineering, **51**, 434–439.

Payne, J.R., McNabb Jr., G.D. and Clayton Jr., J.R., 1991, Oil-weathering Behavior in Arctic Environments, Proceedings of the Pro Mare Symposium on Polar Marine Ecology, Trondheim, May 12–16, Polar Research, **10**, 631–662.

Sayre, W.W. and Chang, F.M., 1969, A Laboratory Investigation of Open-channel Dispersion Processes for Dissolved, Suspended and Floating Dispersants, U.S. Geological Survey, Professional Paper 433-E, pp. 71.

Serrer, M., Sayed, M. and Crookshank, N., 1996, A HYDA Oil Spill Fate Model for the HYFO Decision Support Environment, CHC Technical Report HYD-TR-010. National Research Council Canada, Canadian Hydraulics Centre.

Shen, H.-T., Yapa, P.D., Wang, D.S., and Yang, X.Q., 1993, A Mathematical Model for Oil Slick Transport and Mixing in Rivers, US Army Cold Regions Research and Engineering Laboratory (CRREL), special report 93-21, August 1993.

Spaulding, M.L., 1988, A State-of-the-art Review of Oil Spill Trajectory and Fate Modelling, Oil and Chemical Pollution. **4**, 39–55.

Stiver, W. and Mackay, D., 1984, Evaporation Rates of Spills of Hydrocarbons and Petroleum Mixtures, Environ. Sci. Technol., **18**, 11, 834–840.

# MODELLING OF OIL DROPLET KINETICS UNDER BREAKING WAVES

Z. CHEN[†] & C.S. ZHAN
*Department of Building, Civil and Environmental Engineering, Concordia University, Montreal, Quebec H3G 1M8, Canada*

K. LEE & Z. LI
*Center for Offshore Oil and Gas Environmental Research, Fisheries and Oceans Canada, Dartmouth, Nova Scotia, B2Y 4A2, Canada*

M. BOUFADEL
*Department of Civil and Environmental Engineering, Temple University, Philadelphia, Pennsylvania, 19122, USA*

**Abstract.** A research program was initiated to develop numerical models to describe oil droplet formation and behaviour following oil spills at sea. Specifically, suitable models have been examined to quantify the energy requirement for the formation of oil droplets and their subsequent mixing and resurfacing. Wave energy dissipation rate ($\varepsilon$) is a critical parameter governing the evolution of spilled oil including its droplet size distribution, break-up, coalescence, dispersion, and resurfacing. Four parameters influencing energy dissipation rate were considered, and one was selected for detailed study of oil droplet kinetics including related mixing and transport. Since previous experiments illustrated the evolution of oil droplets maintained under a constant wave energy level, a population model was used to account for the change of droplet size and distribution with time. It is expected that these study results will provide a fundamental understanding of the natural oil dispersion process and the exact nature of the fluid mechanics involved. This information will be used to formulate a standard procedure for testing available and new oil spill countermeasures.

[†] To whom correspondence should be addressed. E-mail: zhichen@bcee.concordia.ca

W. F. Davidson, K. Lee and A. Cogswell (eds.), *Oil Spill Response: A Global Perspective.* 221
© Springer Science + Business Media B.V. 2008

**Keywords:** oil droplet, breaking waves, spills, kinetics, dispersion, wave tank

## 1. Introduction

Oil spills have significant effects on the marine environment. The initial intrusion of spilled surface oil into the water column is caused by agitation from turbulent surface conditions. Wind, waves and currents speed up the shearing of oil droplets, the smallest of which may propel deep into the water, eventually being dispersed by the currents.

The phenomenon of oil droplet vertical mixing has previously been studied through modelling means. For example, earliest models employed the tabulated dispersion rate depending on oil type, sea state, and time after the spill (e.g., Blaikley et al., 1977). Delvigne and Sweeney (1988) and Delvigne and Hulsen (1994) conceptualized an oil droplet entrainment model, elements of which are widely used in oil spill models. To use the oil spill model for a greater variety of oil parameters and environmental conditions, it has to be generalized, utilizing additional theoretical considerations including energy and mass conservation, scale analysis, and the like. Recently, Tkalich and Chan (2002) developed a kinetic model of oil droplets vertical mixing due to breaking waves, considering dominant forces affecting droplet formation and vertical distribution. Laboratory and field verification were also reported. However, previous studies lack thorough examination of the wave energy dissipation rate for determining oil droplet size and the subsequent kinetics under breaking waves. It is of great interest to quantify the energy levels necessary to form and suspend oil droplets in the water column.

Numerous studies on the wave energy dissipation were reported related to beach deformations and the design of coastal structures (Wang and Kraus, 2005). Kaku et al. (2006) conducted a series of studies on hydrodynamics and energy dissipation in rotated flasks for the purpose of testing dispersant effectiveness. For the interests in studying the energy level or requirements, the wave height transformation under irregular breaking waves is an important subject. One of the commonly used concepts to describe energy dissipation is the model proposed by Battjes and Janssen (1978). Their energy dissipation rate model is formulated by combining a single breaking wave with a statistical description of the probability of occurrence of breaking waves. A set of related models of energy dissipation rate have been studied since 1978. Particularly, Rattanapitikon and Karunchintadit (2002) compared seven existing dissipation models for irregular breaking waves using a large number and wide range of wave and bottom topography conditions (total 385 cases from 9 sources of published laboratory data). Their re-calibration and verification studies indicated that the model proposed by Rattanapitikon and Shibayama (1998) gave the best prediction for general cases. By relating the state-of-the-art studies on energy dissipation to the oil droplet kinetics including formation and dispersion through modeling development, improved oil spill models can be developed based on wave tank or field validations.

In addition to the energy requirement and the estimation of oil droplets distribution, oil droplet kinetics includes a combined consideration of droplet vertical dispersion, break-up and coalesce, entrainment and resurfacing, and the related hydrodynamics. The processes can be extremely complicated. An integrated effort based on a full-scale wave tank facility would be desirable to answer fundamental questions before engineering applications. A research program is initiated to develop numerical models to describe oil droplet formation and behaviour following oil spills at sea.

## 2.  Modeling Methods and Approaches

The proposed study includes a systematic consideration of oil droplet kinetics. Firstly, an energy dissipation model will be extended to address the turbulence and energy dissipation in the coastal environment where oil spill occurs. Secondly, the proposed dissipation model will be incorporated to improve models of oil droplets including the approximation of droplet sizes, distributions, and dispersion. Thirdly, turbulent hydrodynamics governs various behaviours of spilled oil in the water column including break-up, coalesce, dispersion, and resurfacing. Attempts are being made to integrate fundamental investigations to develop new oil droplet evolution models under breaking waves in this study.

### 2.1.  ENERGY DISSIPATION RATE

Studies on the energy dissipation rate have been reported. Accurate modeling of the performance of breaking waves was mostly focused on a single breaking wave (Delvigne *et al.*, 1987). There has been a need to model the breaking wave groups and relate to the dispersion of spilled oil. Examination of dissipation models for both regular and irregular breaking waves were reported (e.g., Rattanapitikon and Shibayama, 1998). In this study, a special consideration is made to extend previous efforts on energy dissipation with the unit of Watts/kg, i.e., $m^2 \ s^{-3}$ and an incorporation of a statistical analysis of wave heights:

$$\varepsilon = KQ_b \frac{c_g g}{h H_b} \left[ H_{rms}^2 - \left( h \exp(-0.58 - 2.00 \frac{h}{\sqrt{L H_{rms}}}) \right)^2 \right] \qquad (1)$$

where $\varepsilon$ is the wave energy dissipation rate $(m^2 \ s^{-3})$; $k$ is the coefficient; the published value of $k$ is 0.1; $c_g$ is the wave group velocity $(ms^{-1})$; $h$ is water depth (m); $H_{rms}$ is the root mean square wave height (m), $L$ is wave length (m); and $Q_b$ is the fraction of breaking waves, which is derived based on the assumption that the probability density function of wave height could be modeled with Rayleigh distribution truncated at the breaking wave height, $H_b$ (m).

$$Q_b = -0.738 \left( \frac{H_{rms}}{H_b} \right) - 0.280 \left( \frac{H_{rms}}{H_b} \right)^2 + 1.785 \left( \frac{H_{rms}}{H_b} \right)^3 + 0.235 \ ; \ \text{for} \ \frac{H_{rms}}{H_b} > 0.43 \qquad (2)$$

$$Q_b = 0 \ \text{for} \ \frac{H_{rms}}{H_b} \leq 0.43 \qquad (3)$$

$$H_{rms} = \{ \int_0^\infty h^2 dF(h) \}^{\frac{1}{2}} \qquad (4)$$

where $H_m$ is a maximum possible wave height for each depth $h$ (m) and $F(h)$ is the probability distribution of wave heights, the shape of $F(h)$.

The breaking wave height $(H_b)$ is expressed based on the breaking criteria of Goda (1970):

$$H_b = 0.1 L_0 \left\{ 1 - \exp\left[ -1.5 \frac{\pi h}{L_0} \left( 1 + 15 m^{4/3} \right) \right] \right\} \tag{5}$$

where $m$ is the average bottom slope (rads) and $L_0$ is the deepwater wavelength (m).

## 2.2. WAVE ENERGY

The energy of a wave exists in two forms, potential, due to the deformation of the wave above still-water level, and kinetic, due to the orbital movement of the water particles within the wave form. In the case of the equilibrium energy transfer from the wind to waves and from the waves to whitecaps, the wave energy $E(J)$ will not deviate significantly from the mean wave energy, $E_w$ (J m$^{-2}$). $E_w$ is the wave energy per unit area. The breaking wave energy $E_b$ can be determined using Stokes' second order theory (Michael, 1981):

$$E_b = \frac{g \rho H^2}{16} \left( 1 + \frac{9}{64} \frac{H^2}{k^4 h^6} \right) \tag{6}$$

where $H$ is the significant wave height (m); $\rho$ is the water density (kg m$^{-3}$); $g$ is the gravity acceleration (m s$^{-2}$); and $k$ is the wave number defined by $k = 2\pi/L$.

## 2.3. THE DROPLET SIZE FORMED UNDER BREAKING WAVE

Several models of oil droplets were examined in Tkalich et al. (2002) including the widely accepted Hinze model (Hinze, 1955). The Hinze model is based on the static force balance between the interfacial tension force and the average turbulent pressure forces acting on the maximum stable drop. Neglecting all dynamic effects, the Hinze theory, therefore, does not involve any specific time scale for the drop breakage besides the eddy lifetime (Davis, 1985). The Hinze model does not account explicitly for the influence of the viscosity of the dispersed and continuous phase oil. The Hinze model and its variants can be extended to consider the viscous effects (Davis, 1985):

$$ r_{max} = \frac{c_1}{2\rho^{0.6}\varepsilon^{0.4}} \left( \sigma + \frac{v_0\sqrt{2}(2\varepsilon r_{max})^{1/3}}{4} \right)^{0.6} \tag{7} $$

here, $r_{max}$ is the maximum droplet radius (m); $c_1$ is a constant from 0 to 0.363; $v_o$ and $v$ are the kinematics viscosity of oil and water (m$^2$ s$^{-1}$) for the dispersed phase and the continuous phase, respectively; $\rho$ is water density (kg m$^{-3}$); $\rho_o$ is the oil density (kg m$^{-3}$); $\varepsilon$ is the mean energy dissipation rate based on Equation (1) (m$^2$ s$^{-3}$); and $\sigma$ is the oil-water interfacial surface tension coefficient (N m$^{-1}$).

## 2.4. EVOLUTION OF OIL DROPLETS UNDER BREAKING WAVES

The maximum droplet radius represented by Equation (7) indicates a relatively stable suspension in a continuous phase. The droplet size distribution in an agitated liquid dispersion is a result of simultaneously occurring droplet breakage and coalescence. Pertaining to the time factor, more hydrodynamic effects should be considered. This includes a good simulation of velocity field and energy dissipation rates. The outcome of such effects leads to decreasing of oil droplet size with time for general cases of marine oil spills.

Various studies on oil droplet break-up and coalescence have been widely reported (Tsouris and Tavlarides, 1994). However, it is difficult to incorporate individual studies to the modeling of the oil droplet evolution under breaking waves due to different considerations of energy dissipation and wave hydrodynamic effects. Through an integrated consideration of the related effects particularly involving break-up and coalescence mechanisms, the evolution of oil droplet populations can be represented by:

$$ \frac{dN}{dt} = BN - AN^2 \tag{8} $$

where $N$ is the number of droplets per unit volume; $t$ the mixing time; and $B$ and $A$ are the rate constants for drop break-up and drop coalescence, respectively.

The droplet break-up rate, $B$, is related to a function of its diameter $g(d)$. In the multifractal framework, this function is given by (Baldyga and Podgorska, 1998):

$$g(d) = C_g \left[ Ln \left( \frac{La}{(d)^{1/3}} \right) \right]^{0.5} x_i \phi_i^{1/3} \frac{\varepsilon^{1/3}}{d^{2/9}} \int_{\alpha_{min}=0.12}^{\alpha_x} \left[ \frac{d}{La} \right]^{\frac{2+\alpha-3f(\alpha)}{3}} d\alpha \qquad (9)$$

The terms $C_g$, $x_i$, and $\phi_i$ are dimensionless constants that depend on the system geometry, $d$ is the droplet diameter, $La$ is the largest scale of the system, and $\varepsilon$ is the system-averaged kinetic energy dissipation rate per unit mass. The multifractal representation appears through the terms $\alpha$ and $f(\alpha)$. Smaller values of $\alpha$ indicate larger values of $\varepsilon$. An empirical expression for $f(\alpha)$ was given as:

$$f(\alpha) = a + b.\alpha + c.\alpha^2 + d.\alpha^3 + e.\alpha^4 + f.\alpha^5 + g.\alpha^6 + h.\alpha^7 + i.\alpha^8 \qquad (10)$$

where $a = -3.510$, $b = 18.721$, $c = -55.918$, $d = 120.900$, $e = -162.540$, $f = 131.510$, $g = -62.572$, $h = 16.100$, and $i = -1.7264$. Equation (10) is valid for $\alpha \geq \alpha_{min}$, and $\alpha_{min}$ takes the common value of 0.12.

The drop coalescence rate $A$ is described by Coulaloglou and Tavlarides (1977) as the product of the coalescence efficiency $A(d_m, d_j)$ of droplets of diameter $d_m$, and $d_j$:

$$A(d_m, d_j) = \exp \left[ -K \frac{\mu \rho \varepsilon}{\sigma^2 (1+\phi)^3} \left( \frac{d_m d_j}{d_m + d_j} \right)^4 \right] \qquad (11)$$

where $\mu$, $\rho$ are the continuous phase viscosity and density, respectively; $\sigma$ is the interfacial tension; $\phi$ is the volume fraction of the dispersed phase; and $k$ is the universal constant.

The analytical solution of Equation (8) can be obtained by:

$$N(t) = \frac{1}{A/B + C \exp(-Bt)} \qquad (12)$$

where $C$ is the constant. Thus, the mean droplet size at different times can be obtained through a statistical analysis of the numbers of droplets in different size classes at different times.

Modeling of the evolution of oil droplets can be combined with the consideration of the mixing, diffusion and evaporation processes. Sub-sequently, the concentration of oil in the water column can be predicted.

## 3. A Combined Numerical and Wave-Tank Modeling Study

### 3.1. WAVE TANK

The facility used in this experiment consists of a wave tank as shown schematically in Figure 1. Waves in the tank are generated using a flap-type wave-maker. The conditions of the breaking waves will be created in the tank using the dispersive focusing technique (Longuet-Higgins, 1974). Typically, oil is released onto the surface of the water, the wave maker is started, and dispersant is applied to the slick just prior to the wave breaking. The wave tank measures 32 m long by 2 m deep by 0.6 m wide. It is fabricated from carbon steel and treated with epoxy paint to prevent corrosion. It is equipped with entry and exit ports to accommodate continuous flow simulating ocean currents. A wooden scaffolding structure has been built along the sides of the tank to support deployment of test instruments in the tank and to accommodate technicians.

### 3.2. MODEL INPUT

Table 1 gives the parameters used for the model simulation. In this study, 300 mL crude oil was sprayed into the tank for each test; a dispersant presumably changes the surface tension coefficient from the value of $\sigma = 0.015$ down to 0.002 N/m, the energy dissipation rate from $\varepsilon = 0.2$ to 0.4 m$^2$/s$^3$. The wave conditions currently used are: (1) spilling breaking wave, $H = 12$ cm, $f = 0.85$ followed by $f = 0.48$ Hz; (2) plunging breaking wave, $H = 15$ cm, wave $f = 0.85$ followed by $f = 0.5$ Hz.

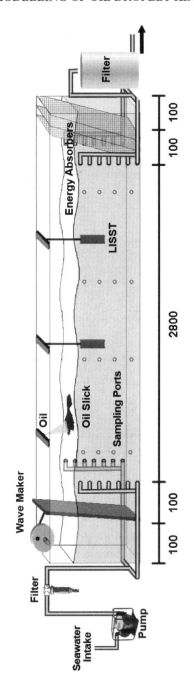

*Figure 1.* Schematic profile of wave tank showing dimensions, placement of wave absorbing screens, wave maker, and holes allowing flow-through to simulate currents (Lee *et al.* 2005).

TABLE 1. The model parameters and values.

| Parameters | Initial setup |
|---|---|
| Water density | $\rho = 1027 kg/m^3$; |
| Oil density | $\rho_o = 878 kg/m^3$; |
| Sea water viscosity | $v = 10^{-6} m^2/s$; |
| Oil viscosity | $v_o = 80 \times 10^{-6} m^2/s$; |
| Minimum radius | $r_{min} = 1 \mu m$; |
| Surface tension coefficient | $\sigma = 0.002 - 0.015 N/m$; |
| Threshold radius | $r_c = 50 \mu m$; |
| The vertical length-scale parameters | $L_d$, $L_{wo}$= 10 m and $L_{ow}$ = 0.2 m; |
| Damping coefficient | $\gamma = 0.0013$; |
| Scaling factor | $\alpha = 1.5$; |
| Diffusion coefficient rate | $D = 0.002 - 0.02$ m$^2$/s; |
| Water depth | 1.25 m; |
| Evaporation rate | $2.0 \times 10^{-5} s^{-1}$; and |
| Coefficient of dissipated wave energy | $k_b$=0.3 |

## 3.3. MODELING RESULTS

### 3.3.1. *Maximum and Mean Radius*

Figure 2 shows the maximum and mean sizes of the resulted oil drop-lets after the spill under a series of wave conditions. Figure 3 gives the similar trend for the formation of mean oil droplet sizes under given wave conditions. Both modeling results are intended to show the energy requirements for the formation of oil droplets.

It is seen that the initial mean radius changes from 140 to 110 μm with energy dissipation rate from 0.2 to 0.4 m$^2$/s$^3$, which also corres-ponds to wave heights from 10 to 20 cm.

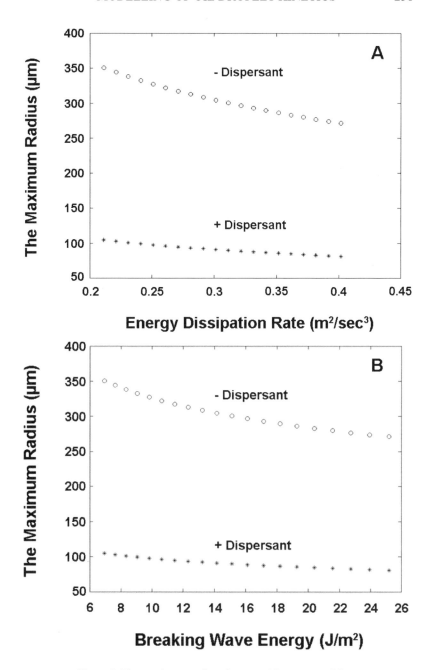

*Figure 2.* The maximum radius changes with wave conditions.

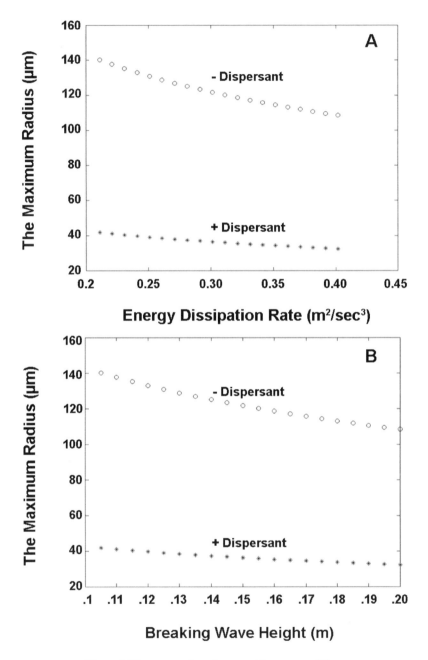

*Figure 3.* The mean radius changes with wave conditions.

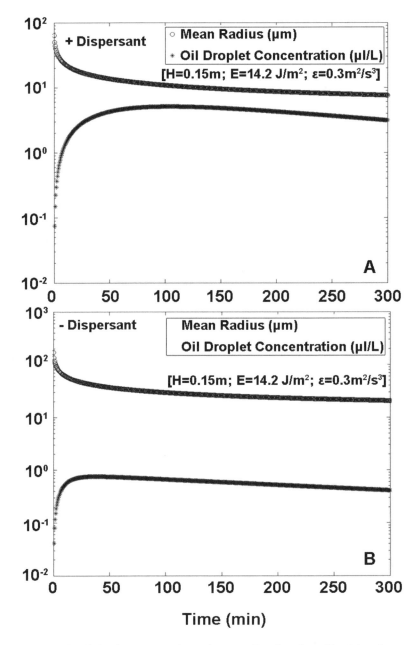

*Figure 4.* The oil droplet concentration and mean radius changing with mixing time.

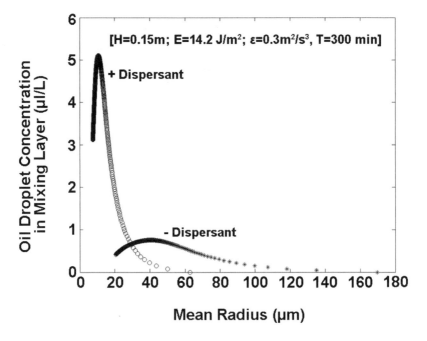

*Figure 5.* The change of oil droplet concentration and mean radius and within the first 300 min.

### 3.3.2. *Evolution of Oil Droplets with Mixing Time*

Taking the 300 mL crude oil as input into the model, Figure 4 shows the results of the mean radius changing with mixing time with and without dispersant. The wave height is 15 cm, wave energy is 14.2 $J/m^2$, and energy dissipation rate is 0.3 $m^2/s^3$. The mean droplet radius decreases from 120 to 20 μm during 300 min without dispersant, at the same time, the oil droplet concentration in mixing layer increases quickly first, then decreases slowly, which indicates that smaller droplet radius will speed the oil entrainment into water column.

During the first 300 min of mixing, the mean radius decreases down to 20 μm without dispersant, but oil droplet concentration in mixing layer increases first to about 1 μl/L and then decreases to 0.4 μl/L, as shown in Figure 5. It is the same trend for the scenario of dispersant application.

## 4.   Concluding Remarks

A research program has been initiated to study the energy and energy requirements for oil droplet formation and the subsequent kinetics after a marine oil spill. This study has included the development of engineering models of wave energy dissipation rate, oil droplet formation, and the droplet evolution with time factor.

The proposed models have been examined through preliminary efforts of wave tank experiments. Reasonable results have been obtained. It indicates that the developed modeling approach can quantitatively address fundamental phenomena and relationships among turbulent breaking waves, the energy requirement, oil and water properties/ interactions, and the resultant oil droplet formation, distribution, and mixing with a time factor.

More coupled numerical and experimental modeling studies using the wave tank facility are ongoing. Future studies will focus on oil droplet fate and transport and the related energy requirement. Effects of oil and water properties and dispersant will also be studied. The results of this research program will provide a fundamental understanding of the natural oil dispersion process and the exact nature of the fluid mechanics involved. A standard procedure for testing available and new oil spill countermeasures can also be formulated based on the combined numerical and physical modeling study.

## References

Baldyga J. and Podgorska, W., 1998, Drop break-up in intermittent turbulence: Maximum stable and transient sizes of drops, *Can. J. Chem. Engng.*, **76**: 456–470.

Battjes, J.A. and Janssen, J.P.F.M., 1978, Energy loss and set-up due to breaking of random waves, *In: Proc. 16th Coastal Engineering Conf., ASCE*, pp. 569–587.

Blaikley, D.R., Dietzel, F.F.L., Glass, A.W., and van Kleef, P.J., 1977, Sliktrak – a computer simulation of offshore oil spills, Cleanup, Effects and Associated Costs. *In: Proceedings of the 1977 Oil Spill Conference*. API, Washington, DC, pp. 45–53.

Coulaloglou, C.A. and Tavlarides, L.L., 1977, Description of Interaction Processes in Agitated Liquid-Liquid Dispersions, *Chem. Eng. Sci.* **32**: 1289.

Davis, J.T., 1985, Drop sizes of emulsions related to turbulent energy dissipation rates. *Chem. Eng. Sci.*, **40**(5): 839–842.

Delvigne, G.A.L. and Sweeney, C.E., 1988, Natural dispersion of oil, *Oil Chem. Pollut.*. **4**: 281–310.

Delvigne, G.A.L., van der Stel, J.A., and Sweeney, C.E., 1987, *Measurement of vertical turbulent dispersion and diffusion of oil droplets and oiled particles*, Report No. Z75-2, Delft Hydraulics Laboratory, Delft, The Netherlands.

Delvigne, G.A.L. and Hulsen, L.J.M., 1994, Simplified laboratory measurements of oil dispersion coefficient – application in computations of natural oil dispersion, *In: Proceedings of the 17th Arctic and Marine Oil Spill Program, Environment Canada*, pp. 173–187.

Goda, Y., 1970. A synthesis of breaking indices, *Trans. Japan Soc. Civil Eng.* **2**: 227–230.

Hinze, J.O., 1955, Fundamentals of the hydrodynamic mechanism of splitting in dispersion processes. *AICHE J.*, **1**: 289–295.

Kaku, V.J., Boufadel, M.C., *et al.*, 2006, Evaluation of mixing energy in laboratory flasks used for dispersant effectiveness testing, *J. Environ Eng., ASCE*, **132**(1): 93–101.

Lee, K., Venosa, A.D., Boufadel, M.C., and Miles, M.S., 2005, *Quality assurance plan for the project wave tank studies on dispersant effectiveness as a function of energy dissipation rate and particle size distribution*, Technical Report, Fisheries and Oceans Canada, Bedford Institute of Oceanography, Marine Environmental Sciences Division.

Longuet-Higgins, M.S., 1974, Breaking waves in deep or shallow water. *In: Proc. 10th Conference on Naval Hydrodynamics*, pp. 597–605.

Michael, E.M. (Eds.), 1981, *Ocean Wave Energy Conversion*. Wiley, New York.

Rattanapitikon, W. and Karunchintadit, R., 2002, Comparison of dissipation models for irregular breaking waves, *Songklanakarin J. Sci. Technol.*, **24**(1): 139–148.

Rattanapitikon, W. and Shibayama, T., 1998, Energy Dissipation Model for Irregular Breaking Waves, *In: Proceedings of the 26th Coastal Engineering Conf.*, ASCE, pp. 112–125.

Tkalich, P. and Chan, E.S., 2002, Vertical mixing of oil droplets by breaking waves, *Marine Pollut. Bull.*, **44**(11): 1219–1229.

Tsouris, C. and Tavlarides, L.L., 1994, Breakage and coalescence models for drops in turbulent dispersion. *AICHE J.*, **40**(3): 395–406.

Wang, P. and Kraus, N.C., 2005, Beach profile equilibrium and patterns of wave decay and energy dissipation across the surf zone elucidated in a large-scale laboratory experiment, *J. Coastal Res.*, **21**(3): 522–534.

# THE NECESSITY OF APPLYING SAR IMAGERY TO OIL SPILL MODELING IN CASES OF DATA OBFUSCATION

M. PERKOVIC[†] & L. DELGADO
*University of Ljubljana, Faculty of Maritime Studies and Transportation, Pot pomorscakov 4, SI-6320 Portoroz, Slovenia*

M. DAVID, S. PETELIN & R. HARSH
*Transas Techologies., Maly pr VO 54/4, 199178 St. Petersburg, Russia*

**Abstract.** Oil spill monitoring is all the more important during circumstances that prevent immediate action, such as the recent crisis in Lebanon during the Israel-Lebanon war. Conditions of war prevented the acquisition of routine and necessary information, such as: the precise quantity of oil lost to the sea, the rate and duration of escape flow (in this case initial reports were that the oil was 'spilled' during two separate events, yet subsequent images suggested continuous flow over a period of at least two weeks), the type of oil, precise locations, and shore characteristics. Accurate simulation of the Lebanon slick was possible only by using SAR imagery which, for one instance, demonstrated that the behavior of the slick ran counter to expectations informed by knowledge of winds, currents, and waves; that is, though the main mass behaved according to models, running up along the coast was a thin layer of oil spread up to 20 km out to sea, extending northward approximately 200 km. If this event may be of any benefit, it will be that it leads to the improvement of the applicable models by considering what has been learned by running our models for an extended period over a vast area in these particular circumstances.

**Keywords:** oil spill monitoring, modeling, SAR imagery

---

[†] To whom correspondence should be addressed. E-mail: Marko.Perkovic@fpp.uni-lj.si

W. F. Davidson, K. Lee and A. Cogswell (eds.), *Oil Spill Response: A Global Perspective.* 237
© Springer Science + Business Media B.V. 2008

## 1.  Introduction

"In the course of the conflict in the Middle East, the oil-fuelled power plant south of Beirut was hit by bombs in the middle of July. Part of the storage tanks caught fire and were burning for several days. A large part of the heavy fuel oil emitted into the Mediterranean Sea as a result of the blast". That was the first telephone message from the Oil Spill Team Joint Research Centre of the European Commission on the last day of July. "Are you ready to participate in the simulation of the likely spread of the oil slick within these holidays?" An explanation followed that satellite photos would be processed shortly that would be an additional source of information for the start of simulation, along with information from the Lebanese Ministry of Environment (MoE). And so at the Faculty of Maritime Studies, which functions under the University of Ljubljana, the Maritime Training and Crisis Management System Simulator Center (MCSC) found itself in a real oil spill crisis situation. Previously, it was used for identifying ships that release illicit oil or oily water into the North Atlantic, training seafarers and stake-holders, and improving contingency plans.

## 2.  Materials and Methods

For any other complex prediction of the oil slick spread, oceanographic and meteorological conditions of the local area are needed from the start. In the Lebanon case, where the oil slick spread up to 200 km along the coast, it is not possible to accurately foretell events without the use of programs for simulating drift fate and weathering of oil slicks. For instance, in a case like this with the oil slick traveling along the coast, with which it is constantly in contact, stranding reduces the floating part of the mass and the exposure to breaking waves and tide currents can wash the stranded slick back out to sea. How long the slick will stay or travel with the current and wind depends on the discharged quantity, oil type, and conditions in the field. The mass of the dischar-ged oil dec-reases immediately upon release due to evaporation of volatile com-ponents. Some fuels evaporate completely in hours. Rough seas, high wind speeds and high temperature will further increase the rate of eva-poration. The process of evaporation increases the density of the re-maining floating mass, which allows for separation and the potential for sinking. This phenomenon is typical for heavy fuel oils that already have a high density, and during the night when it is cooler these oils can exceed the density of water and thus sink (ITOPF, 1987).[1] A slick discharged on a sandy beach can mix with the sand and then sink when washed to sea. Many oils, especially those with asphalt content grater than 0.5%, tend to absorb water to form water-in-oil emulsion which can increase the volume of pollutant by a factor of between three and four. Such emulsions are often extremely viscous; thus, processes which would typically dissipate the oil are retarded. The emulsion can partially sink, but mostly this is the main reason for the persistence of light and medium crude oils on the sea surface. Waves and turbulence on the sea surface may also act on the slick to produce oil droplets with a wide range and size. This weathering process, known as natural dispersion, additionally reduces the mass of the floating slick. Small droplets that remain in suspension become mixed into water columns. But on the other hand, this process may increase the polluted area by forming a very thin film; i.e., large droplets rise back to the surface behind the advancing slick.

The discharge in Lebanon is especially typical because it has spread out over a great surface area which, as the wind field was constantly changing, so did the sea currents that pushed the slick out to the sea, back to the coast and North along the coast. The coastline is variegated, from sandy beaches, rocky tracts, gravel, pebbles, to individual infra-structures such as fishing ports, marinas and beaches where a great quantity of oil landed and remains. Because of insufficient information regarding the quantity and duration of the discharge and chemical con-tent of the oil, an accurate simulation is not possible without the use of an additional source of information, in this case satellite images.

From the satellite images the exact quantity of the discharged oil, or the quantity of the stranded oil, on the coast cannot be determined, but it is possible to predict future spreading and so warn the neigh-boring countries of the potential danger in time.

---

[1] Over a 10°C temperature range the density of sea water will only change by 0.25% where oil density changes by 0.5%.

## 2.1. MARITIME TRAINING AND CRISIS MANAGEMENT SYSTEM SIMULATOR CENTER

As part of preparing the National Oil and Chemical Spill Contingency Plan for Slovenia (NOCSCP), in which the Faculty of Maritime Studies and Transportation has taken active part, the Potential Incident Simulation, Control and Evaluation System is a response simulator intended for preparing and conducting command centre exercises and area drills in oil spill response. The Potential Incident Simulation, Control and Evaluation System (PISCES) is oriented to accomplish tasks required by the Oil Pollution Act 1990 (OPA 90) to provide improved training for spill response managers. PISCES provides the exercise participants with an interactive information environment based on the mathematical modeling of an oil spill interacting with surroundings and combat facilities. The program uses vector-based nautical charts to provide a geographical description of the incident area. Vessel traffic and positions of response resources are shown on the chart. The system also includes information-collecting facilities to enable the assessment of the participants' performance. A typical view of the PISCES user interface is shown in Figure 1, where the user defines the incident area, including the outline and type of the coastline that may be affected by the spill. The type of coastline (e.g., "rock", "mud", "sand", etc.) determines the maximum amount of stranded oil and oil beaching ratio. Environmentally sensitive areas are set as a polygonal object which can detect the presence of pollution and generate alarms.

The most important environmental factor is the field of currents. It is defined via a set of "current vectors" similar to that denoted on nautical charts. These vectors may be entered manually based on the information from tidal and current tables, or imported from files in a simple text format provided by local sources. Other environmental conditions include: wind speed and direction, wave height, water and air temperatures. Time variations of the field of currents and weather parameters are defined as tabular dependences for the entire duration of the scenario. For the description of oil pollution, the location of spill source and physical characteristics of the spilled oil product can be specified. The type of source may be instant release or moving leakage, representing, for example, a drifting damaged tanker. Multiple spill sources may be set in one scenario. Finally, a preliminary inventory is taken of

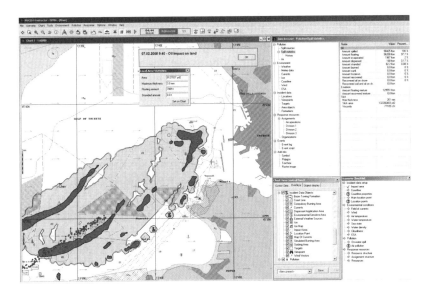

*Figure 1.* Typical view of PISCES user interface.

response resources that can be determined and used in the scenario and assigned to on-scene locations.

### 2.1.1.  *Oil Spill Modeling*

The PISCES model reflects the common understanding of the behavior of oil spills on the sea surface. The model describes oil spills on the water surface in the first days of release when response operations are carried out. The initial amount of spilled oil varies from 10 to thousands of tons. For the description of an oil slick the PISCES model uses the Lagrangian approach, which is an effective technique of oil spill numerical simulation. The oil spill is represented by an ensemble of particles moving under the effect of wind and current. The distinctive feature of the model is the extension of the Lagrangian approach by introducing interactions between oil particles (Delgado *et al.*, 2005). This innovation allows compensation for some essential deficits of the traditional Lagrangian methods and describes, in a satisfactory manner, oil interaction with different kinds of natural and artificial barriers.

    The processes occurring in the oil slick are also taken into account
and include evaporation, natural dispersion, emulsification, and viscosity
variations. Simulations are carried out with regard to the environment
representation, including complex coastline descriptions and a variable
field of currents and weather conditions. In addition, models of response
operations such as burning, booming and skimming are implemented.
The movement of an oil slick is assumed to be two-dimensional and
resulting from transport by currents and wind, spreading and stochastic
diffusion. According to generally accepted approximation, oil particles
move 100% with a stream and 3% downwind. Diffusion adds to this
movement a random component. The spreading of the oil slick is
adjusted to match the well known approximation of Fay (1971), with
corrections introduced by Lehr *et al.* (1984) to take into account downwind elongation.

    For the description of weathering processes, oil product is rep-
resented in the same way as in the ADIOS (NOAA, 2000) oil library
and characterized by a common product name, type (e.g., crude or
refined), specific gravity, surface tension, viscosity, distillation curve,
emulsification constant, pour and flash points. Evaporation process is
simulated by using a pseudo-component approach and uses the evapo-
rative algorithm proposed by Stiver and Mackay (1984) and Mackay
*et al.* (1980). This method calculates evaporation rate as a function
of wind, slick area and oil properties like the distillation curve. Processes of emulsification and natural dispersion are described with
empirical dependencies proposed by Mackay. Both dependencies take
into account strength of wind and oil properties. The emulsification process starts as soon as the fraction of evaporated oil reaches the emulsification constant, which is determined from the experimental data.
Emulsifi-cation stops when water content in the floating emulsion
mixture rea-ches its maximum. The maximum is an empirical value. It
is usually about 75% for crude oils and 25% for refined products. The
algorithm for natural dispersion takes into account oil-water surface
tension, oil viscosity and slick thickness. Viscosity variation due to
emulsion formation is defined by the Mooney (Mooney, 1951) equation, and the evaporation effect is taken into account according to the
Mackey solution.

    Presence of ice has a considerable effect on oil transport and spread-
ing as well as the weathering processes. It is also necessary to take into

account oil encapsulation into ice. This part of the model has been designed primarily for illustrating qualitatively the main behavior features of an oil spill in icy conditions due to the complexity of the problem. Ice is presented as a set of polygons defined manually by the operator. The user specifies the shape and location of ice regions, ice thickness and age. For pack ice, the user additionally specifies ice concentration, which is understood as the ratio of the area covered by ice to the whole area of the pack ice region. While the air temperature is decreasing from 20°C to the Pour Point value, the oil viscosity gradually increases and spreading speed decreases. Oil spreading and weathering processes stop when air temperature reaches the Pour Point value. Within the pack ice area, oil moves along currents and wind with a speed slower than in open water. Spreading and evaporation is also lesser than in open sea. The decrease of movement, spreading and evaporation speeds depends on ice concentration. Within a fast ice area, evaporation stops because oil is assumed to move under the ice. The oil moves and spreads under fast ice with a speed lesser than in open water. The decreasing of movement and spreading speed depend on ice thickness. The thicker the ice, the slower is the speed. Oil thickness also affects spreading. The spreading stops when oil thickness reaches the critical value.

Besides PISCES, the 3D oil spill model MEDSLICK (www. oceanography.ucy.ac.cy/cycofos/medslik-act.html) is used (more intensively during this study). MEDSLIK incorporates the evaporation, emulsification, viscosity changes, dispersion in water column, and adhesion to coast. The oil spill is modeled using a Monte Carlo method. The pollutant is divided into a large number of Lagrangian parcels of equal size. At each time step, each parcel is given a convective and a diffusive displacement. Emulsification is also simulated, and the viscosity changes of the oil are computed according to the amounts of emulsification and evaporation of the oil. MEDSLICK is able to read meteorological and oceanographic data ADRICOSM (www.bo.ingv.it/adricosm) for the Adriatic Sea as well the Levantine Basin (where the pollution in this case occurred) and some other areas of the Mediter-ranean Sea. For Levant atmospheric and oceanographic SKIRON (3 h time-steps) wind and high (2.5 km) resolution (6 h time-step) currents and wind CYCOFOS (www.oceanography.ucy.ac.cy/cycofos/forecast. html) forecasting products could be used. (Cyprus ocean forecasting system.)

### 2.1.2. *Configuration of Slovenian Crisis Center*

The configuration of the customized installed crisis equipment is quite unique. The PISCES simulator is connected to three ship handling simulators, of which one is Full Mission on a floating platform. Besides the navigational simulator, which has Vessel Traffic Component (VTS), and Search and Rescue (SAR) module with helicopter console is integrated with eight communicating GMDSS stations. There is a full mission engine room simulator which is also connected to the ship handling simulator. A cargo handling simulator (CHS), as part of the ERS simulator, is located on the main bridge. The PISCES simulator is also equipped with five operative stations, designed for Civil Protection, Port Authority, Marine Police, Navy and Researchers from the Faculty or the Environmental Agency. The simulator can work not only as educational equipment, but also for actual assistance on the occasion of an oil spill (Perkovic *et al.*, 2006). PISCES can receive on-line data from the Automatic Identification System receiver, which is located on Mt. Slavnik and directed by the Port Authority. Because of the ideal height of the receiving antenna, it is possible to supervise ships as far as the city of Split (200 nautical miles) (Perkovic, 2002).

Besides the abovementioned, traffic data is acquired and relayed to the simulator center with RADAR based VTS located at the marine police station. Meteorological and oceanographic parameters are integrated into the GIS on near real time from the marine biological station.

Chemical processes in the combustion of the stain are also presented on PISCES GIS. The process is calculated with ALOHA software made by the NOA – National Oceanic and Atmospheric Administration. The simulator center is further equipped with an AIS receiver, Marine VHF transceiver, HF/MF receiver and digital decoder, phones and fax machine. Detailed structure of the Crisis System is described in Figure 2.

### 2.2. RELEVANT INFORMATION ON THE OIL POLLUTION ALONG THE COAST OF LEBANON

The initial simulation was based on information announced by the EC Monitoring and Information Centre (MIC) (http://ec.europa.eu/environment/civil/leb_cy_2006.htm) on the 28th of July. The day before, Lebanon requested urgent help to contain the environmental damage

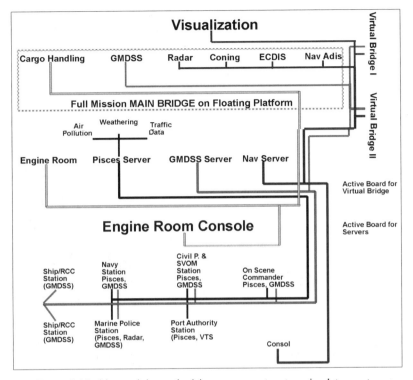

*Figure 2.* Maritime training and crisis management system simulator center.

caused by a major oil spill off its coast north of Beirut, including large amounts of dispersants, booms, absorbents, skimmers as well as specialized vessels and equipment needed to clean water of pollutants. According to the Lebanese authorities' first message from 21 July, the spill was caused by the destruction of a storage tank at a power plant at Jieh, which was hit by Israeli bombs on the 13th and 15th of July. The fuel had been burning, and an amount of it spilled into the sea.

The first report from the Lebanese authorities, dated 21 July 2006, indicated that 10,000 t of oil escaped from the storage unit and the potential total could eventually reach 35,000 t. Based on information acquired from the source of the spill, power plant storage tanks, in the first simulation it was expected that the discharge was heavy oil IFO 380. Wind and current information were obtained from the average daily forecast of the CYCOFOS (Figures 3 and 4).

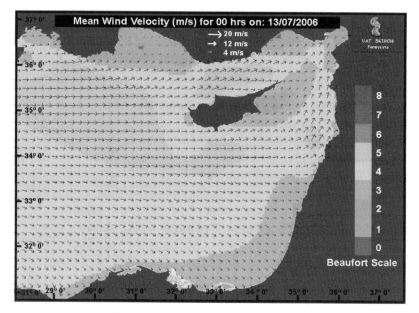

*Figure 3.* SKIRON wind fields on July 13.

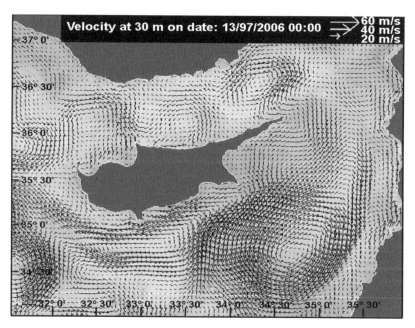

*Figure 4.* CYCOFOS sea currents.

The second simulation was already based on the use of more precise information obtained from REMPEC »info n.1« (http://www. rempec.org/news/asp) declaring that on the 21 July 2006, the affected area was reported to be located 20–30 km north of the source, and on 24 July 70–80 km north of the source (one third of Lebanese coastline). These areas are composed of sandy beaches, rocky beaches, fishing ports and marinas. The precise quantity of discharged oil was still unknown. REMPEC liaised with its network of experts, providing the first estimate of the potential behavior of the oil, on the basis of the information on oil characteristics sent by the Lebanese MoE. The product can be compared to IFO 150 (medium/heavy oil).

Following a request for help from the Lebanese Ministry of Environment, the EC Monitoring and Information Centre (MIC) of DG Environment triggered the International Charter 'Space and Major Disasters' in order to obtain information about the extent of the oil pollution along the coastal strip and, where possible, the extent of the pollution. The Center for Satellite Based Crisis Information of DLR (Germany's national research centre for aeronautics and space) produced the first MODIS map provided by NASA on August 3, and the first SAR (Envisat ASAR) image provided by ESA on August 5.

The Joint Research Center of the European Commission was activated at the same time. Already on the same day, July 30, the oil spill team created the first map from the low resolution Synthetic Aperture Radar frame obtained by ESA, and on the next day a high resolution SAR image from the 21st of July was released. Important information was gained from both SAR images (http://oilspills.jrc.it/ Lebanon) showing that the oil did not travel entirely along the coast, rather that some of the slick was carried out to sea, which was totally against expectations, based on the northerly current and predominant southwesterly winds.

In the first days of August, numerous satellite images were processed. An image taken on August 3 (http://oilspills.jrc.it/Lebanon) confirmed the presence of the slick in the Gulf of Tarabulus. The Syrian authorities confirmed the presence of oil slicks on the sea and by the coast. By the 6th of August the oil slick had spread up to 200 km and it was clearly visible from high resolution SAR images that the oil was still discharging in moderate quantities in the region of the power plant of Jieh (http://oilspills.jrc.it/Lebanon). The spread of the slick on this day was towards the north-east, but three days later, on the 9th of August, it was clear that there was an extensive discharge to the north.

On August 19, Envisat ASAR Wide Swath was acquired and delivered to JRC by ESA. A zoom in over Jieh and Beirut showed some dark features that have been identified as oil slicks (JRC Info N. 7) (http://oilspills.jrc.it/Lebanon). In particular, there was an area that indicated that oil was still present along the Lebanese coast from Jieh up to southern Beirut. The extent of this spilled area is about 20 km long and 2 km wide. A second, larger spill area was identified further north. The extent of this second spill was about 30 km long and 5 km wide, located along the coast from northern Beirut up to Batroun. It is important to note that, in comparison with the preceding observations (JRC Info Reports), the overall extent of the spilled areas along the Lebanese coasts was reduced.

On the 22nd and 25th of August, small oil patches could be identified on SAR images (JRC Info N.8 and N.9). On the SAR image from the 29th of August the patches cannot be seen.

The final simulation was based on corrected entry information. The program was upgraded with an additional function which can produce foot prints of the oil slick movement. The major work was done on the developing code for reading the CYCOFOS data format, which enables the entry of the currents and wind field of the whole Levantine basin.

## 3. Results

### 3.1. INITIAL SIMULATION RESULTS

Right after the start of the first simulation it was clear that PISCES software is not designed, at the moment, for such wide oil pollution analysis where there is an enormous area of an oil slick connected to the coast; because the program is basically meant for training with crisis situations on the sea, it has to calculate progress in very small time steps (15 s). Such small steps are necessary to appropriately simulate the dynamic response of deployed resources. Because of these small steps a two week simulation was carried out in just 12 h. As the first simulation finished (Figure 5a-e), the first satellite image was acquired by MODIS on NASA's AQUA satellite (Figure 6) on the 24th of July and was used in the report of the research team from the

Oceanography Centre of the University of Cyprus (Zodiatis *et al.*, 2006). Right after that the SAR image (Figure 7) was available in a lower resolution, on which the state of the oil slick can be seen for the 21st of July.

Both pictures did confirm the reconciliation of the transport between the "real" (Figure 7) and the simulated (Figure 5a–e) state. The SAR picture (Figure 7) additionally reveals facts that before were not known. From the picture it is clearly visible that circular currents were

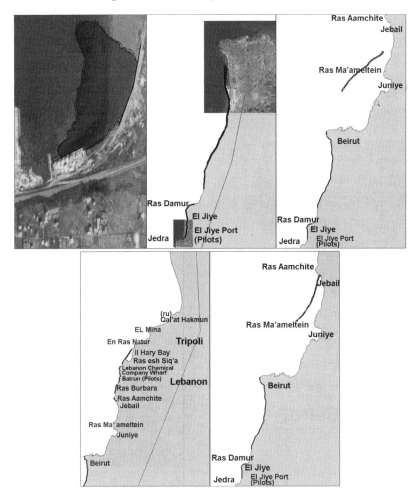

*Figure 5 (a–e).* PISCES simulation results.

forming beyond the Beirut peninsula that further dispersed the oil slick, which spread throughout the gulf between Beirut and Jebail.

The most important information obtained from the SAR picture is the actual range of the discharge that is entirely contradictory to the report that MoE sent to the REMPEC center (see Section 2.2 for the second simulation). The oil slick that was thought to be restricted to the outskirts of Beirut, had in fact already spread 30 miles north (Figure 7).

The oil spill weathering results were, on the other hand, very bad. In the first case, a very simple coastline was used, with no expressive points that would capture a part of the slick. A rocky coast was defined that could retain at most 10 t of oil per longitudinal kilometer. Because of a chance mistake in the entry of data, the waves were much higher than in reality, resulting in a dispersion of 27% and distorting the quantity of emulsion. On the other hand, the loss of mass

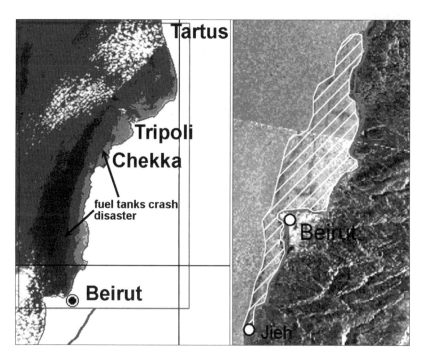

*Figure 6. Modis-Aqua (NASA) July 24, 10:55 GMT.*

*Figure 7. Envisat Asar (Esa) July 21, 07:50 GMT.*

because of evaporation was greatly underestimated, only 2% of the slick evaporated. Because of the low stranding and great emulsion, the slick had a tendency to persist on sea for a long time which means that it could pollute as far as Syria and even Turkey and Cyprus.

## 3.2. DEBRIEFS AND CORRECTIONS

Because of bad results from stranding and evaporation, a more accurate electronic map was prepared before the second simulation; a greater part of the coast as sand tracts with an absorption capacity up to 100 t per longitudinal kilometer was defined. At the same time, some estimations of evaporation process for a big spill of IFO 180 oil were made. Figure 8 describes the simulation of the first five days of 15,000 m$^3$ spill when IFO 180 (14°API) was used on water at 25°C and wind speed 10 m/s. The PISCES model took into account the slick area, which has a noticeable effect on the dynamic of the evaporation process.

Three approximations of slick area were used from (Figure 9): fixed area (assuming 1mm slick thickness), calculated according to Fay and Mackay formulas. As can be seen, the Fay and Mackay approximations provide much lower values of evaporation. This is quite understandable

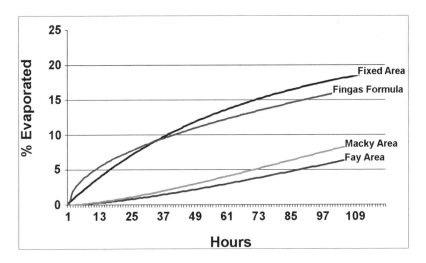

*Figure 8.* Evaporation curves.

because of the enormous size of the spill. The area calculated with Fay and Mackay formulas cannot reach the fixed area size even after 120 h. The spill created in accordance with the satellite image from 21 July has an area (of "thick oil") about 100 km$^2$, which is much greater than the Fay and Mackay predictions. PISCES uses the Mackay approximation of a spill area, which is the main reason it underestimates evaporation in big spills of heavy oils. The Fingas approximation (green) virtually conforms to the fixed area graph (dark blue) suggesting, taking into account the results of satellite observations, its tendency to yield more accurate results.

The next important goal was to use not just the average wind field and sea currents, but to import high resolution SKIRON winds and CYCOFOS currents. Some programming work was necessary to convert the data to the appropriate "xml" format. The current version of PISCES is able to use only a single wind vector, which is suitable only for small affected areas. If the area is large, as in Lebanon, it is possible to use the main wind value at the position where the main part of the oil slick is located. In accordance with the expected transport of

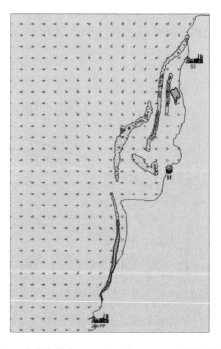

*Figure 9.* Oil slick approximation area on July 21st.

the slick it is possible to arrange a wind table. The latter method is not fully applicable in this case because of the continuous discharge in the south and the simultaneous spread of the slick for distances of up to 100 nautical miles, wind directions varying.

## 3.3. FINAL SIMULATIONS

In the last simulation, corrected entry data was used. A continual oil spill was simulated from the 15th of July to the 10th of August. First, a forecast of currents and winds from the CYCOFOS base (time span 12 h) was used. The simulation yielded very surprising results. The oil slick remained south of Beirut. In the continuation, a simulation followed on the CYCOFOS (time span 6 h) base of currents in com-

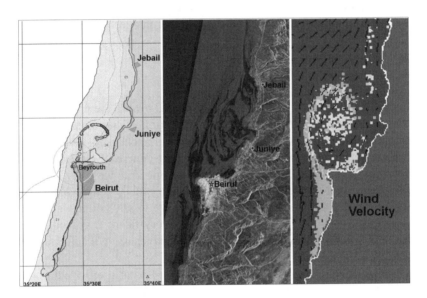

*Figure 10.* (Left) PISCES simulation results for CYCOFOS 6 h forecast surface currents and wind analyses.

*Figure 11. (Middle)* SAR image on 6th of August.

*Figure 12.* (Right) MEDSLICK simulation results for CYCOFOS 6 h forecast, 30 m depth currents.

bination with the SKIRON winds (3 h). In this example, the acquired result was the same. A simulation with the same parameters was performed with the use of the program package MEDSLICK and the acquired result was the same. The simulation published by Zodiatis presented a completely different course of dispersion of the oil slick. They have the ability to use MEDSLICK with SKIRON wind generated in time span 1 h. In the continuation, the deviation was checked between the SKIRON 3 and 1 h in such a way that the slick was simulated only with the wind by using PISCES. No great difference was noticed. A much better result was achieved with the integration of currents and wind analysis obtained from ECMWF – European Centre for Medium-Range Weather Forecasts. The foot print simulation is shown in Figure 10. Spill statistics demonstrate that 17.5% of the oil evaporated, 74.2% stranded and 7.8% sunk. Interestingly, another simulation performed using the current field at 30 m depth provided results most similar to the real case suggested by the SAR images.

## 4.  Conclusion

Perhaps more important than what was learned toward the improvement of oil spill simulation is the further illustration of the damage to the environment when a spill is allowed to spread unchecked for an extended period of time, whatever the reason. Of course, the condition of war may be given as the cause, but at the same time we know that the spill was certainly going to spread northwards, and thus there was never the slightest danger of environmental damage to any country south of Lebanon. This must be stressed because every factor involved, from data entry to assessment of simulation effectiveness, was affected by the delay in the relay of satellite imagery.

Nonetheless, all scientists and service did what they could under the circumstances. In the case of the PISCES simulation, the most important factor was that SAR imagery was an invaluable aid in assessing the process. In fact, changes have already been made that enhance the accuracy of our crisis simulation capacity. Further, during the process it was verified that the more collaboration with other institutions, the more efficiently and accurately results may be obtained.

## 5. Acknowledgements

We would like to acknowledge and thank the EC JRC oil spill team and Zodiatis *et al.* for sharing their resources.

## References

ADRICOSM – Adriatic Sea Integrated coastal areas and river basin management system pilot project, http://www.bo.ingv.it/adricosm/

CYCOFOS Bulletin, http://www.oceanography.ucy.ac.cy/ cycofos/forecast.html

Delgado, L., Kumzerova, E., Martynov, M., Mirnyj, K., Shepelev, P., Ed: C.A. Brebbia (Eds) and M. de Conceicao Cunha (Eds), 2005, *Dynamic Simulation of Marine Oil Spills and Response Operations.* Coastal Engineering VII. Modelling, Measurements, Engineering and Management of Seas and Coastal Regions, WIT Press, pp. 123–133

Fay, J.A., 1971, Physical processes in the spread of oil on a water surface. *Proc. on Prevention and Control of Oil Spill*, American Petroleum Institute, Washington, DC, pp. 463–467.

ITOPF, 1987, *Response to Marine Oil Spills.* The international tanker oil pollution federation, Witherby, London.

Lebanon and Cyprus emergencies, 2006, European Commision, Civil Protection, Monitoring and Information Centre (MIC), http://ec.europa.eu/environment/civil/ leb_cy_2006.htm

Lehr, W.J., Cekirge, H.M., Fraga, R.J. and Belen, M.S., 1984, Empirical Studies of the Spreading of Oil Spills. *Oil Petrochem.l Poll*, **2**, 7–11.

Mackay, D., Buist, I.A., Marcarenhas, R. and Paterson, S., 1980, Oil spill processes and models. *Environment Canada Manuscript Report No. EE-8.* Ottawa, Ontario.

MEDSLICK Oil Spill Application, http://www.oceanography.ucy.ac.cy/cycofos/ medslik-act.html

Mooney, M., 1951, The viscosity of a concentrated suspension of spherical particles. *J. Colloid. Sci.*, **10**, 162–170.

NOAA, 2000, *ADIOSTM (Automated Data Inquiry for Oil Spills) version2.0.,* Hazardous Materials Response and Assessment Division, Prepared for the U.S. Coast Guard Research and Development Center.

Oil Spill at the Lebanese Coast, Center for Satellite Based Crisis Information, DLR, http://www.zki.caf.dlr.de/applications/2006/lebanon/lebanon_2006_en.html

Oil spill off the coast of Lebanon, International Charter "Space and Major Disasters", http://www.disasterscharter.org/disasters/CALLID_126_e.html

Perkovic, M., 2002, AIS and Global Safety. 10th scientific-technical symposium ISEP, Electro-technical Society of Slovenia, 3–4 October 2002, Ljubljana.

Perkovic, M., Suban, V., Petelin, S., David, M., 2006, Complex maritime system simulator centre for advance training and education. V.J. Bezjak (Ed.). *4th International Science Symposium*, 19–21 April 2006, Portoroz, Slovenia. Ljubljana 2006, pp. 835–839.

Spill in Lebanon, REMPEC, Sitrep 1-16, http://www.rempec.org/news.asp. Oil pollution in Lebanon, European Commission Joint Research Center, Info 1-10, http://oilspills.jrc.it/lebanon

Stiver, W. and Mackay, D., 1984, Evaporation rate of spills of hydrocarbons and petroleum mixtures. *Environ. Sci. Tech.*, **18**, 834–840.

Zodiatis, G., Lardner, R., Hayes, D. and Georgioul, G., n.d., The Cyprus Coastal Ocean Forecasting and Observing System, a contribution to GMES Marine services, http://www.oceanography.ucy.ac.cy/cycofos

Zodiatis, G., Lardner, R., Hayes, D., Soloviev, D., and Georgioul, G., 2006, *Oil Spill Modeling Predictions in the Mediterranean, Lebanon Costal Oil Spill Pollution,* http://www.oceanography.ucy.ac.cy/cycofos

# PART 5. RISK ASSESSMENT/CONTINGENCY PLANNING/OPERATIONAL RESPONSE/POLICY

# PREVENTION OF OIL SPILLS CAUSED BY PREMEDITATED NEGATIVE INFLUENCE

V. KRIVILEV[†]

*Academy for Geopolitical Problems, 5-ay Magistrinya Str. 10/2, off. 21, Moscow, 123007, Russia*

**Abstract.** The increase in terrorist activities during recent years makes it necessary to take special measures of counteraction. This threat also poses a risk for the oil and gas complex. In this presentation, special attention was paid to the definition of the specific character of premeditated and unpremeditated acts increasing the risks of oil spills. With the purpose of prevention, localization and liquidation of oil spills, we consider the question of defining critical areas of concern where aggressive actions may lead to emergency situations. The specificity of production, transportation and processing of hydrocarbon raw materials is also considered. A separate section of the presentation is devoted to the development and application of the of informational cooperation system between countries/participants for the liquidation of oil slicks during premeditated aggressive acts against the environment, food chains and water systems. The prototype of such a system is being developed within the framework of the Russia-NATO Council "Ecoter" Project by specialists from Russia, Italy, Romania and others.

**Keywords:** oil spill response, terrorism, Ecoter, counteraction

---

[†] To whom correspondence should be addressed. E-mail: vkrivilev@sops.ru

W. F. Davidson, K. Lee and A. Cogswell (eds.), *Oil Spill Response: A Global Perspective.*   259
© Springer Science + Business Media B.V. 2008

# SPILLVIEW: A SUPPORT TO DECISION-MAKING SOFTWARE IN EMERGENCY RESPONSE TO MARINE OIL SPILL

M. BLOUIN[†]

*Canadian Coast Guard, Department of Fisheries and Oceans Canada, Quebec Region, 101 Champlain Boulevard, Quebec City, Quebec, G1K 7Y7, Canada*

**Abstract.** In the event of a marine oil spill, it is necessary to quickly and clearly assess the situation and estimate the extent of the area potentially impacted by oil. This software combines the following features integrated in a Geographical Information System: geo-referenced digital aerial survey; access to trajectory forecast model results; and, charts with marine and terrestrial data. These features allow for better planning of emergency response in terms of deployment of personnel and equipment, because it helps to clearly document the observed spill while subsequently providing the length of the coastline at risk and the forecasted time for which the oil spill will reach the coast. Aerial surveys are one of the main tools used towards these ends. Aerial observations support the planning of oil cleanup and recovery work, and can also provide accurate data for oil spill fate and trajectory models. Aerial surveyors traditionally use paper maps to record their observations. This method presents some limitations. These include: (1) the difficulty of evaluating the exact location of observed features on the map; (2) the difficulty of recording all the necessary information on a fixed-scale map; and (3) the issue of transferring the recorded observations to spill managers, which takes time, requires explanations from the observer and can be subject to miss-interpretation. It is for these reasons why the Canadian Coast Guard, in partnership with Cogeni Technologie Inc., developed the SpillView software system. SpillView, which runs under the Windows XP operating system, is designed to operate on a pressure sensitive tablet PC equipped with GPS and electronic maps. The system displays the real time location and trajectory of the air-craft. The observer can record different types of observations (e.g., oil location, environmental resources, and shorelines contamination) on

---

† To whom correspondence should be addressed. E-mail: blouinm@dfo-mpo.gc.ca

W. F. Davidson, K. Lee and A. Cogswell (eds.), *Oil Spill Response: A Global Perspective.*    261
© Springer Science + Business Media B.V. 2008

georeferenced layers that can be individually exported to formats compatible with other Geographical Information Systems. The observer can also use the system to electronically transfer the observed oil location to a spill modeling center, and display the modeling results within minutes. Spillview proved to be a good tool to support training and exercises, as it can be used to portray different spill scenarios on electronic maps. The software could also be utilized for other aerial survey needs, such as national security or forest fires. SpillView is presently being enhanced to provide operational support by enabling real time access to equipment inventory databases and fieldwork description forms. Following a positive response from CCG and their partner, Cogéni decided to develop a commercial version of the system able to meet the high expectations of the international market. A serious initiative was undertaken to develop software and to set up the necessary infrastructure required to support customers during the product's integration and throughout its lifespan.

**Keywords:** SpillView, oil spill response, modeling, coast guard

## 1.  Introduction

One of the roles of the Canadian Coast Guard (CCG), a branch of the Department of Fisheries and Oceans (DFO), is to act as the lead agency in the event of a marine spill occurring in Canadian waters. As defined in the "Marine Spills Contingency Plan National Contingency Chapter" (Government of Canada, 2004), the CCG also has a mandate to ensure response preparedness in the event of an environmental emergency. In addition to the provision of training, organization of spill simulation exercises, and development of partnerships with the private sector, the CCG contributes to the development of response methods and tools that will improve the efficiency of a response. One of these development efforts, financed within CCG's National Research and Development program, resulted in the creation of the SpillView software.

## 2.  Context

One of the first steps following a marine oil spill is to clearly assess the situation and estimate the extent of the area potentially impacted by oil. The Quebec region includes more than 13,700 km of coastline, mostly located in remote and sparsely populated areas. In these conditions, aerial reconnaissance is the most efficient method to capture the data necessary to make early response decisions. Once captured, this data must be processed, analyzed and transmitted to the decision makers involved in the response.

Currently, aerial survey observations are recorded during the flight on paper maps or sketches. This way of doing things presents a number of limitations. These include: (1) the difficulty for the surveyor to evaluate the exact location of observed features on the map; (2) the difficulty to record all the necessary information on a fixed-scale map; and (3) the issue of transferring the recorded observations to spill managers, which takes time, requires explanations from the observer and can be subject to miss-interpretation.

To solve these issues, it was decided to develop a computerized system that takes advantage of recent developments in data capture technologies. These include: the availability of small and powerful tablet computers; the development of wireless digital data transmission networks (cell phones, satellites, etc.); the Global Positioning System (GPS); and the availability of georeferenced electronic maps. The new system was to support the capture of information used to plan oil recovery and shoreline treatment, and also to gather real time input data for an oil spill dispersion model. This would allow better predictions of the movement of oil slicks.

## 3. System Design, Structure and Functions

The data capture software, called SpillView, is a C/C++ application developed by Cogeni Technologie (Quebec, Canada). The system is designed to work on a tablet PC equipped with a pressure sensitive monitor (Figure 1). SpillView can also be operated on a standard Windows-based computer using any of the following operating systems: Windows 2000 and XP. The system can capture and display cartographic data in the ESRI Shapefiles (SHP).

The SpillView software can use topographic maps or nautical charts as base maps. It can read the signal from a GPS linked to the serial port of the computer, and use the information to display the position of the aircraft and capture data. SpillView supports two general types of tasks: data capture of oiling observations and spill modeling.

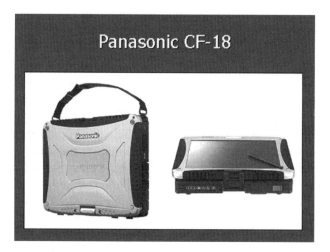

*Figure 1.* Hardware platform used to operate the SpillView system during aerial reconnaissance.

## 3.1. CAPTURE OF AERIAL OBSERVATIONS

When interacting with the software, the user is presented with a topographic map or nautical chart centered on the position of the air-craft. As the aircraft moves, the cartographic display is automatically re-centered. The system automatically records a trace of the aircraft movement, showing the flight path. This trace can be saved and replayed at will. The observer can use the tactile monitor to record a number of features, such as the location of an oil slick, or the presence of oil along the shoreline. The data can be recorded by using pre-set standard categories in order to reduce the necessity to "key in" values, since the tablet computer does not have a keyboard. The main functions are available through buttons and menus. The system also includes tools typically found in Geographical Information Systems (GIS), enabling distance measurements, zooming, displaying pre-defined views, etc.

The main georeferenced data capture functions include:

- The description of oil slicks according to their tar code, following a method adapted from the oil slick characterization method developed by Allen and Dale (1995). The function evaluates the oiled surface area and provides a rough approximation of the oil volume (Figure 2).

- A qualitative description of shoreline oiling, following four categories (none, light, medium and heavy oiling). Shoreline oiling observations are displayed following a simple color code. The function also provides an estimation of the length of oiled shoreline by oiling category (Figure 3).

- A description of sensitive environmental resources (birds, marine mammals, etc.). The system can capture the species name, number of individuals observed, and a description of the situation.

Once captured, the data can be directly transmitted to the decision makers at the command center through any network supporting the transfer of digital data, such as cellular or satellite networks. Only captured data layers are transmitted. This ensures that the size of the transferred files remains small.

### 3.2. SPILL TRAJECTORY MODELING FORECAST

A special set of functions provide SpillView with the capacity to obtain a forecasted trajectory of the oil spill using the results of an ocean model. The model is operated daily by DFO Ocean Sciences Branch at the Maurice Lamontagne Institute.

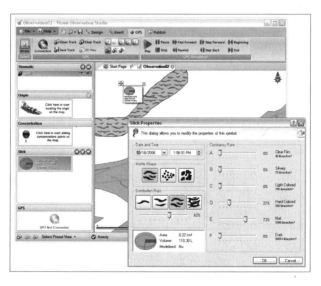

*Figure 2.* Portion of the SpillView user interface showing an oil slick, captured by the operator, and a dialog window used to describe oiling characteristics.

The model domain covers the Gulf of St. Lawrence. The ocean boundaries are Belle-Isles and Cabot straits. The upstream boundary is set on the St. Lawrence River at Trois-Rivières. Twice a day on the modelling server, the forecasted wind from the operational atmospheric model of the Meteorological Service of Canada is received. Using this forecast along with the tidal forcing and the freshwater runoff, an ocean model is run and issues a forecast of ocean currents for the next 48 h. The hourly wind from the MSC observing stations is also received. The server is linked through DFO internal network between the CCG in Quebec City and the ocean model server in Mont-Joli. A process is running in the background to keep it on a waiting mode for a connection to the Spillview interface. Two ocean models are used: one for the whole Gulf of St. Lawrence on a 5 km grid and one for the upper St. Lawrence on a 400 m one.

When a trajectory simulation is initiated through the SpillView interface, the following fields are available: date and hour at the beginning of the trajectory, and in what time zone; forward or backtrack trajectory up to 36 h; position of the observed slick. This observation

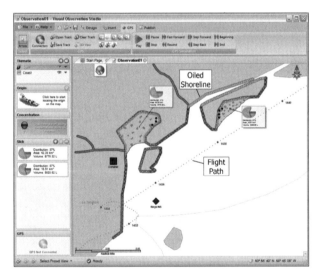

*Figure 3.* Portion of the SpillView user interface showing oiled shorelines and oiling categories. To record an observation, the operator first selects an oiling level, and then uses a pointer to follow the correspondingly oiled section of the shoreline on the pressure sensitive monitor.

can be entered either in latitude and longitude, or by drawing its contours on the map from an aerial observation. The mean trajectory of the movement of the slick will be displayed as well as the contour of the patch at selected times.

The trajectory of the oil spill is calculated on the server in Mont-Joli by adding its displacement generated first by the surface currents of the ocean forecast model and second by the direct action of the wind using a factor of 3.3% of the wind intensity in the downwind direction. Observed winds closest to the incident are used and merged into forecasted winds when needed. Prior to the calculation, the winds that will be used are shown for validation and edition. Common sense, other available data, consultation with a local forecaster, can all be input to contribute to this validation process.

The spreading of the oil spill is reproduced by using on the order of a hundred initial points with horizontal mixing randomly applied. The evolution of the mean trajectory and the evolution of the oil spill at specified times are returned to the interface to be displayed over the map of the area. The limitations of the modelled trajectory near the coastline are illustrated by displaying the coastline seen by the model, which is quite irregular at a 5 km resolution.

The nature of the oil spilled is not taken in account in the modelling. Since the initial phase of spreading, during which the nature of the product makes a difference, is short, we neglect it. Rapidly, the spreading of the oil is governed by the movement of the surface waters which is independent of the nature of the spilled product. If the hydrocarbons are below the surface, such as Orimulsion, the movement is governed only by the currents in the water column and can be obtained by specifying so in the interface.

Before the development of SpillView, an emergency officer needed to contact an expert from the DFO modeling group to obtain trajectory forecasts. This could introduce a delay of 1 h and up to 3 h when the incident was after working hours. The SpillView system uses its communication capabilities to automatically access the modeling server and obtain the forecasted trajectory in 2–3 min (Figure 4). Since the bulk of the calculation is done remotely, a limited quantity of data needs to be transmitted:

From the PC of the response officer, a series of latitude and longitude positions to define the contour of the spill.

1. From the server and back: A series of hourly wind intensity and direction.

2. From the server: A series of forecasted latitude and longitude positions to define the contour of the spill.

### 3.3. OTHER FUNCTIONS

The software can import and display information layers provided in ESRI Shapefiles (SHP) format files. The system can also be used to create and attach mission related information files and describe the position of the field response team. Research tools are also available, enabling the creation of search templates and exclusion zones. All the information is displayed as layers on top of the base-map, and each of these layers can be exported as an ASCII file following the MID/ MIF format. SpillView stores all information in one single file called "Patrouille", in its own format. Information within this file can be

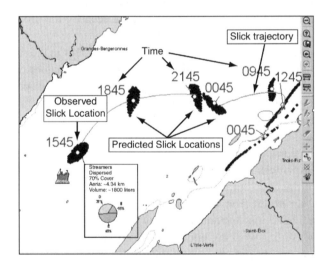

*Figure 4.* SpillView Map window displaying the forecasts of the spill model. The forecast uses the location of an oil slick entered by the user as a starting point. The model provides the slick trajectory (line), as well as the spreading of the oil (black points) at different time intervals.

extracted or exported, used to create thematic maps, or simply display all available information. Thematic maps can be saved as standard bit-maps in GIF, JPG, BMP or PNG format. The system can also produce a textual report (in .RTF format) providing a summary of each of the themes. Printing options include the possibility to select individual layers, add a title, explanatory comments, and a warning message. In addition, it is possible to zoom and pan maps in print-preview mode.

## 4. Discussion

SpillView is now fully functional. Nonetheless, improvements are already planned. An interface that will receive and display real time data transmitted from a drifter buoy developed by the CCG is currently being developed. The buoy is designed to follow oil slicks and transmit its position as well as the air and water temperature through the Orbcomm satellite network. When available, this function will provide SpillView with the capacity to support the validation and enhancement of the spill modeling forecasts. In addition, real time data transmitted from the buoy will make it easier to localize the slick in the morning, always a difficult task after a night a-drift without positive observation.

Additional tools will also be developed to exchange information with our partners through Environment Canada's Internet-based Génie-Web system. This system was developed to integrate all of the data captured by government organizations involved in an incident. Providing direct access to databases of response equipment or bird inventories is considered.

SpillView proved to be an excellent tool when used for exercises and training. The system could also prove very useful to support aerial reconnaissance surveys in domains other than spill response, such as aerial surveillance of pollution, to validate the position of vessels or in forest fire delineation.

## 5. For More Information

For the latest information on SpillView, visit Cogéni Technologie world Wide Web site at: ttp://www.cogeni.com/Products/VOS/VOS.en.aspx

# References

Allen, A.A. and Dale, D.H., 1995, Oil Slick Classification: A System for the Charac-
    terization and Documentation of Oil Slicks, *In: Proceedings International Oil
    Spill Conference*, American Petroleum, Washington, DC, pp. 315–322.
Government of Canada, Department of Fisheries and Oceans, Canadian Coast Guard,
    2004, *Marine Spills Contingency Plan National Contingency Chapter*, Home
    page: http://www.ccg-gcc.gc.ca/rser-ssie/er-ie/plan/main_e.htm

# ACCIDENTAL MARINE POLLUTION IN BELGIUM: THE EMERGENCE OF RESPONSE STRATEGIES

T.G. JACQUES[†] & F. DELBEKE
*Unité de Gestion du Modèle Mathématique Mer du Nord, (UGMM/MUMM) Institut Royal des Sciences Naturelles de Belgique, Gulledelle, 100 B-1200 Bruxelles, Belgium*

**Abstract**. Belgium is a small country bordering the southern North Sea near one of the busiest shipping routes in the world. Marine management in general and government involvement in environmental emergencies at sea in particular have evolved tremendously over the past 25 years. Changes were at first stimulated by the difficulties and often confusion encountered in dealing with shipping accidents, such as the loss of the Mont-Louis in 1984 and the tragedy of the ferry Herald of Free Enterprise in 1987. From a defensive and rather passive attitude in the 1980s, government services moved through a decade of building up awareness to a more alert and co-operative level of preparedness. Government scientists having an interest in marine management were very influential during this transition, which lead to a comprehensive packet of legislation. Since the turn of the millennium, the Administration's attitude has become overtly offensive, supported by European law which lays a strict liability on the polluter. An elementary Coast Guard structure has been established, equipment has been purchased, and operational plans have been adopted. How these measures will succeed in mitigating the impact of marine pollution remains to be demonstrated.

**Keywords:** marine pollution, oil spill response, policy

---

[†] To whom correspondence should be addressed. E-mail: t.jacques@mumm.ac.be

W. F. Davidson, K. Lee and A. Cogswell (eds.), *Oil Spill Response: A Global Perspective.* 271
© Springer Science + Business Media B.V. 2008

## 1.  Introduction

Strategies for dealing with accidental marine pollution run over an
entire array of situations and measures, from warning protocols to compensation claim procedures, and preside over contingency planning. At
whichever level it develops, a response strategy does not stand alone. It
fits a particular administrative structure, complies with international law
and national legal requirements, adapts to a given distribution of powers
and depends on budgetary constraints. More directly, it responds to
prevailing risks in a specific geographic setting, and seeks to protect
exposed resources. In developing their response strategies, coastal States
have certainly taken inspiration from internationally accepted doctrine,
but they have had to accommodate their own particular context, and
learn from their own experiences.

Marine management in general and government involvement in
environmental emergencies in particular, have evolved considerably
over the past 25 years. A case in point is Belgium, where response arrangements have shaped up over time and are in the process of further
evolving. We suggest that this evolution reflects a profound change in
mentalities and attitudes toward both shipping and the environment in
that same period of time. Context and perception have shaped the strategy.

## 2.  Setting the Scene: The Belgian Coast at Risk

Belgium is a small industrialized country bordering the Southern
Bight of the North Sea. The country has only 65 km of coastline, most
of it a sandy beach extending in a straight line along a high energy
shore facing the north-west. On the landside, the beach is fringed with
dunes, both recent and ancient, most of them having been subjected to
development for recreational housing. The waterfront is heavily built
and the beaches in summertime are a major tourist resort. This shoreline is cut through at only four places by narrow channels: the Yser
river in Nieuwpoort, the Ostend harbor, a small yachting harbor in
Blankenberge, and the Zeebrugge seaport from where a sea canal
connects to Bruges, 12 km inland. Zeebrugge has expanded seaward
into a vast outer harbor and natural-gas terminal. At the border with
the Netherlands, the beach opens into an interesting salt marsh that is

carefully managed as a nature preserve. A similar environment, only much smaller, is found on the right bank of the Yser estuary. Both natural habitats are highly vulnerable to oil pollution. The coastal waters are visited by large numbers of birds (Haelters *et al.*, 2004). In wintertime, they include Guillemot, Little Auk, Common Scoter, Little Gull and Great Crested Grebe. In summertime, terns brood in the Zeebrugge outer harbor and on the shore. Three marine protected areas have been established under European Union legislation for the Sandwich Tern (*Sterna sandvicensis*), the Common Tern (*Sterna hirundo*), the Little Gull (*Larus minutus*) and the Great Crested Grebe (*Podiceps cristatus*). The Little Tern (*Sterna albifrons*) also broods in Zeebrugge where it enjoys a special protection.

At the eastern end of the Belgian coast, the French border is only a few miles from Dunkirk and the Channel entrance. This is one of the busiest ship transit areas in the world. Vessels of all kinds follow the IMO traffic separation scheme en route to or from Rotterdam, Hamburg and other North Sea and Baltic ports, while ferries sail across. Fishing vessels and pleasure boats abound. Norcontrol IT (2006), working under contract of the UK's Maritime Coastguard Agency (MCA), reports 93,509 ship movements through the Dover Strait in 2000. Thirty nine percent of those vessels were carrying some hazardous cargo. Just before entering Belgium's exclusive economic zone, part of the east-bound traffic branches off towards the Scheldt estuary (see Figure 1). There are more than 37,000 arrivals in Belgian ports per year, and Anatec (2003), cited by Germanischer Lloyd Offshore and Industrial Services (2003), estimated that shipping in the Scheldt river mouth amounted to 62,588 movements a year.

As these numbers indicate, the risk of shipping accidents in the vicinity of the Belgian coast is potentially high. The 12-mile zone is characterized by a system of elongated, shallow sandbanks fanning out in a north-easterly direction. The combination of shoals, frequent foul weather and ocean mist when conditions are calm make navigation risky. Beaconed channels must be followed scrupulously. VTS (Vessel Traffic System) monitoring is a must. Pilotage services are offered in the approaches (North Sea pilots) and in the Scheldt estuary (Scheldt pilots).

In spite of this dynamic seascape, the shortness of her coastline never allowed Belgium to become a true seafaring country. Aside from a highly committed fishing community and trade interests revolving around the harbors, the Belgian establishment has not over the past, as a rule, paid much attention to marine management. Interestingly, much of the awareness of the potential of marine resources and the need to protect them has been rooted in the scientific community. The department of Science Policy has run oceanographic research programs involving research institutions and all large universities since the early 1970s. Such efforts often got the enthusiastic support of the small – but very alert – Belgian Navy. Until the early nineties, counter-pollution preparedness was entirely based on the good will of the military and initiatives of Science Policy. How this general disposition and the unfolding of events since 1980 succeeded in raising the general awareness of the need for more formal arrangements can be illustrated with a few remarkable incidents that have stood out in that quarter of a century.

## 3.  The Trial of the Eighties

From 1981 to 1989, nineteen accident alarms were reported to the coastal station of Ostend (Table 1). Not all those events took place in Belgian waters and not all of them developed into a casualty, but all were felt to threaten Belgian interests sufficiently to justify spreading the alarm and calling for some inter-departmental consultation. Ten of these incidents involved a risk of oil pollution. For the other nine, the presence of dangerous cargo caused concern. The response to such alarms was often chaotic. On August 25, 1984, the French Ro-Ro (Roll-on/Roll-off) ship Mont-Louis collided with the car ferry Olau Britannia. After drifting some hours, the Mont-Louis sank only a few cables off the three-mile limit of Belgium's territorial sea. The ship had sailed from Le Havre with thirty 15-t cylinders of uranium hexafluoride on board. $UF_6$ is a product of nuclear fuel reprocessing. It is not very radioactive but extremely corrosive. It took 20 days for the Belgian authorities to obtain the ship stowage plan and to get a satisfactory confirmation of the number of cylinders, their tonnage, and the nature and properties of $UF_6$ (Jacques, 1985). Meanwhile, the French authorities had sent divers to the wreck right after the sinking, as it was their right on a vessel flying their flag in the international waters. The ship owner con-

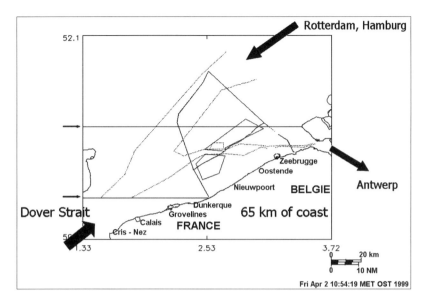

*Figure 1.* The southern part of the North Sea showing Belgium's coast and zone of jurisdiction, the zone of joint responsibility of Belgium, France and the United Kingdom under the Bonn Agreement (between the two horizontal arrows), and the IMO traffic separation scheme.

tracted a salvage company to recover the cylinders. This situation caused considerable tensions between the responsible services of the two countries, Belgium considering that it was under threat and France putting the lid on the information because anything relating to nuclear products was considered sensitive. There was reluctance to make full use of the official channels of information established under the Bonn Agreement (Bonn Agreement, 1983). As a consequence, information was exchanged simultaneously through diplomatic channels, between responsible government services directly, and at a political level. The content of the information did not always agree.

It took 40 days for the salvage company to recover all $UF_6$ cylinders. The Belgian authorities had put up an elementary monitoring program for seawater contamination and run plume models to figure out the safety risks if a recovered cylinder leaked $UF_6$ in the air. By the end of the operation the two countries had come to good understanding and full co-operation (Figure 2). The entire dangerous cargo

TABLE 1. Alarm incidents having required inter-departmental consultation in the eighties, named after the vessel causing the greatest concern.

| Year | Vessel | Source of concern |
|------|--------|-------------------|
| 1981 | World Dignity | 100,000 t crude oil |
| 1982 | Jumpa | Toxic containers |
| 1982 | Saint Anthony | 38,000 t crude oil |
| 1982 | Molesta | Fuel oil |
| 1982 | Benetank | 3,000 t heavy fuel oil |
| 1983 | Sterling | 55,000 t crude oil |
| 1984 | Mont-Louis | Nuclear re-processed material |
| 1985 | Stamy | Fuel oil |
| 1985 | Contract Voyager | Dangerous drums |
| 1986 | Staffortshire | Light petroleum gas |
| 1987 | Herald of Free Enterprise | 5 Dangerous lorries |
| 1987 | Olympic Dream | 2,100 t gasoline |
| 1987 | Skyron | 137,000 t crude oil |
| 1988 | Borcea | Fuel oil |
| 1988 | Seafreight Fairway | 3 dangerous lorries |
| 1988 | Anna Broere | 550 t acrylonitrile |
| 1988 | Westeral | Dangerous containers |
| 1989 | Paul Robeson | Grounding |
| 1989 | Perintis | 5.8 t lindane |

was recovered with no harm to personnel or the environment. But in Ostend, decisions had been made in a crisis committee under sometimes very confusing circumstances. A single meeting held under extreme media pressure had been chaired successively by the Governor, the Minister of the Environment and the Prime Minister himself. Orders were followed by counter-orders. No mitigation strategy had been agreed upon in case of pollution resulting from fuel loss. The Belgian authorities largely failed to establish a sound coordination with the salvage company and the ship owner. They in fact had little control of anything until all valuables had been removed from the wreck.

Three years later, in the evening of March 6, 1987, the British car ferry *Herald of Free Enterprise* of Towsend Thoresen took water through the main garage opening on leaving Zeebrugge, and in a few Three years later, in the evening of March 6, 1987, the British car

*Figure 2.* The wreck of the Mont-Louis in 1984. International experts inspect the casualty after the dangerous cargo has been removed.

ferry *Herald of Free Enterprise* of Towsend Thoresen took water through the main garage opening on leaving Zeebrugge, and in a few minutes capsized in shallow water near the entrance channel. Not less than 150 passengers and 38 crew members lost their lives while attempting to escape (Department of Transport, 1987). During that night, about 350 people were rescued and brought to safety by boats and military helicopters in the most dramatic maritime rescue operation in Belgian history. Twelve hours after the accident, a first list of dangerous goods carried on board was made available to the authorities (Jacques, 1990). It turned out that five lorries contained more than 100 different chemicals, most of them in very small quantities. The most worrisome were 25 drums of toluene di-isocyanate and six drums of wastes and hardening salts containing cyanide. Five tons of hydroquinone, a product commonly used in photography but toxic to marine life, was reported to be on board.

In spite of the tragic death toll, the rescue operation had been carried out with skill, good sense and courage under the co-ordination of the Governor of West Flanders. The salvage of the vessel and cargo was going to require long efforts and sophisticated means, but it was kept under reasonable control with the same spirit and dedication. Channels of communication were opened between the maritime authorities, the government marine scientists responsible for environmental monitoring and the consortium of salvage companies. Water contamination was controlled in and out of the vessel. This did not prevent the ship owner, however, to go to court against the State out of fear that the authorities would interfere with the salvage of the vessel. All decisions relating to environmental protection subsequently had to be made under court supervision. The owner eventually accepted to deploy limited oil containment equipment to intercept possible leakages. Although less than half the cargo had been recovered by the end of the salvage operation on April 27, there was no evidence of environmental damage.

As the decade drew to a close, the awareness of the requirements and pitfalls of large salvage operations had built up in those government circles that were most directly facing the responsibility to deal with marine incidents. The salvage industry had been recognized as an inevitable stakeholder and a possible ally. The technical, legal and financial implications of counter-pollution had been recognized. The need for prevention was pressing. In October 1987, Belgium extended its territorial limits to 12 miles. On November 25, 1987, a first agreement for aerial surveillance of ship-borne pollution was signed with the Netherlands, Denmark and Germany. On March 11, 1988, the Cabinet approved the North Sea Disaster Plan, a general framework for contingency planning that is still in effect today. On September 22, 1989, all North Sea countries adopted amendments to the Bonn Agreement extending co-operation to airborne surveillance of the entire basin. Yet, at that point, the attitude of the authorities in dealing with pollution threats had slowly moved from passive to reactive. They were still pretty much putting up a defensive stand.

## 4.  The Formative Nineties

The following decade was not spared its share of emergencies. No less than 29 shipping accidents caused the alarm to spread through the early warning system (Table 2). Some were spectacular. In 1990 the chemical tanker Bussewitz ran aground in the Scheldt, not far from Antwerp, with 14,000 t of ammonia. In 1992, the Long Lin collided with the Amer Fuji in the French zone and light oil pollution contaminated much

TABLE 2. Alarm incidents reported through the early-warning system in the nineties.

| Year | Vessel | Source of concern |
|------|--------|-------------------|
| 1990 | Bussewitz | 14,000 t ammonia |
| 1990 | Thomas Weber | 221 Dangerous drums |
| 1990 | Viva | Fuel oil |
| 1991 | Tomisi | Fuel oil |
| 1991 | Globel Ling | Fuel oil |
| 1991 | Clipper Confidence | Lead concentrate, Cu, Zn |
| 1991 | Grete Turkol | Ethylbenzene |
| 1991 | British Esk | Naphtha |
| 1992 | Jostelle | Fuel oil |
| 1992 | Atlantic Carrier | Fuel oil |
| 1992 | Nordfrakt | 3,252 t lead sulphide |
| 1992 | Long Lin | Fuel oil |
| 1992 | Amer Fuji | Fuel oil |
| 1993 | Fleur de Lys | Fuel oil |
| 1993 | Alexandros | Fuel oil |
| 1993 | Zaphos | 68,000 t condensate? |
| 1993 | Sherbro | Bags pesticides |
| 1993 | Hyaz | Fuel oil |
| 1993 | British Trent | 24,000 t gasoline |
| 1993 | Aya | Fuel oil |
| 1994 | Shoeburyness | 90 M14 mines |
| 1994 | Elatma | 1,378 t $NH_4 NO_3$ |
| 1994 | Ming Fortune | 38 t sodium chlorate |
| 1995 | Carina | Fuel oil |
| 1995 | Spauwer | Capsizal, oil |
| 1997 | Bona Fulmar | 7,000 t gasoline lost |
| 1997 | Mundial Car | Oil |
| 1997 | Vigdis Knutsen | Oil tanker |
| 1999 | Ever Decent | Toxic containers |

of the Belgian coast. In 1993, the Western Winner collided with the tanker British Trent in the international waters of Belgium's continental shelf and the tanker, laden with gasoline, caught fire, causing the loss of several lives.

Belgium's maritime community in the nineties was still suffering from a deep depression and it was in a grim mood: the entire merchant fleet had passed under Luxemburg's flag, the fishing fleet was drastically reduced, the Navy was cutting its personnel by half, and the State-owned ferry company RMT (Régie des Tranports Maritimes) was facing grave financial difficulties. Marine pollution was widely perceived as an increasing, unavoidable threat. In this context, the entry into force in 1994 of the United Nations Convention on the Law of the Sea (UNCLOS, 1982) and the long ratification process, completed by Belgium in 1998 served as a healthy stimulus (Jacques, 1995). Under the impulse of the departments of Foreign Affairs and Science Policy, treaties were signed with France (October 8, 1990), the United Kingdom (May 29, 1991), and the Netherlands (December 18, 1996) to delimit Belgium's continental shelf and the future exclusive economic zones (EEZ). Between 1990 and 1995, the North Sea States engaged into consultations to agree on a common legal content for the EEZs. In 1999, Belgium's EEZ was established by law, granting the State jurisdiction on environmental matters.

Meanwhile, the authorities attempted to improve their preparedness for dealing with accidental and operational ship discharges. The Management Unit of the North Sea Mathematical Models (MUMM), a federal scientific department, and the department of Defense were appointed to develop means of aerial surveillance for the North Sea. Regular surveillance started in 1991 with twin-engine propeller aircraft. On the initiative of the oil industry, the International Tanker Owner Pollution Federation wrote a country study presenting recommenddations to Belgium for contingency planning (ITOPF, 1994). The same year, MUMM obtained from the federal department of the Environment a mobilization contract with a salvage company, the Towage and Salvage Union (Union de Remorquage et de Sauvetage, URS), for an oil combating vessel equipped with two Lori oil recovery systems and dispersant spraying booms (Figure 3). That contract remained active until June 2001. In 1994 again, the State obtained the first compensation for pollution damage in the Borcea claim, for a spill that had occurred in 1988. In 1997, a first conviction was obtained for an illegal oil discharge recorded by an observation aircraft. Over the entire period, regular training courses and exercises where organized jointly by MUMM, the Navy and the Civil Defense.

*Figure 3.* Multipurpose vessel equipped with oil-recovery systems under contract of
the Belgian authorities (1994–2001).

It is clear from the above that the authorities had learned from their
experience and were slowly getting organized. When the containership
Samia was pulled clear of the refrigerator ship Carina with which she
had collided in 1995 (Figure 4), she left the latter with a big hole in
the engine room and an oil leakage. MUMM's personnel representing
the State environmental authorities accompanied the Navy On Scene
Commander on board of the casualty to direct containment and remediation measures, and remained there until the ship was safely back in
port. When in 1999 the Ever Decent asked permission to enter a
Belgian port for unloading after a fire had raged through several con-
tainers in UK waters, threatening to release hazardous substances, permission was only granted after a combined visit of Belgian maritime
and environmental authorities on board, in full co-ordination with the
British authorities.

In 1999, a bill submitted by the Cabinet was adopted by Parliament
and became the Law on the Protection of the Marine Environment in
the Marine Spaces under Belgium's Jurisdiction (MMM, 1999). The
law sets a solid framework for marine management, including a permit
regime for economic activities at sea, and protected areas. It confirms

the strict liability of polluters and grants the State all necessary powers to intervene at sea, enforce counter-pollution legislation, and impose combating and remediation measures; thus, at the turn of the millennium it can be said that, from defensive, the attitude of the authorities had become active and alert. The driving force behind the progress had largely been Science Policy. The theme had been to test the organization and the techniques, to train personnel, and to provide a strong legal ground for a sound strategy in counter-pollution response.

*Figure 4.* Collision between the containership Samia and the refrigerator ship Carina in Belgium's zone in 1995.

## 5. The Aggressive 2000s: From Defense to Attack

Aside from the environmental management as such, it had been felt after the adoption of the 1999 MMM law that the spaces under the new jurisdiction of the State required a structural co-ordination of all legally responsible departments, both federal and local (the latter being placed under the authority of the regional Flemish Government). Long consultations and the patient work of several inter-departmental working groups led to the creation of a Coast Guard Structure. The structure remains purely administrative and has received no new means, whether material or budgetary. Nonetheless, it now provides a platform for better communication, regular consultation, and mutual assistance between partners. A co-operation agreement formalizing these arrangements was signed by the Flemish and federal governments in 2005.

The developments of the previous two decades had not resulted in any new capacity for combating oil at sea with State-owned means. No investments had been made (Table 3). In year 2000, the federal Department of the Environment, a department of the former Ministry of Health, was allocated about 2.5 million euros to invest in equipment. Enough booms, skimmers and spraying gear were acquired to equip two strike teams at sea and to open 10 small working sites along the shore. This equipment has been deployed on several occasions, either in training or in real response operations. Unfortunately, no dedicated vessel has been acquired or contracted yet, and the open-sea strike teams must be assembled by calling on vessels of opportunity. As the table shows, the budgets allocated to working costs remain prohibitively low.

Now that a legal framework for counter-pollution response existed and that the need for means and structure had been recognized, the Administration felt that it could move ahead. In 2002, the federal Cabinet decided to entrust one Minister with the co-ordination of North Sea matters. It also created a small administrative cell for marine environmental protection within the federal Department of the Environment, a move that was badly needed and had been delayed only too long.

TABLE 3. Allocations in Belgium's federal budget for counter-pollution response in the North Sea and on the shore (EURO).

| Year | For operations | For investments |
|---|---|---|
| Yearly, 1994–1999 | 74,000 | 0 |
| 2000 | 72,000 | 2,479,000 |
| 2001 | 72,000 | 0 |
| 2002 | 112,000 | 0 |
| 2003 | 686,000 | 0 |
| 2004 | 278,000 | 15,000 |
| 2005 | 231,000 | 71,000 |
| 2006 | 303,000 | 72,000 |

Such was the situation in December 2002, when the car ferry Tricolor sank off Dunkirk, close to the Belgian zone, after a collision. Lying in shallow water, the wreck was hit by several other passing vessels, causing repeated oil pollution. The French and Belgian authorities agreed on a permanent on site watch and the Belgian Navy engaged several vessels in the operation. France ordered the wreck to be removed. It was cut in sections by specially equipped salvage pontoons, in an operation that lasted until September of the next year. Oil discharges during that operation were never massive enough to permit an efficient recovery at sea, but the Belgian coast was oiled repeatedly and government strike teams were mobilized several times, with a view to testing the new equipment. Several full-scale beach cleanup operations took place under co-ordination of the Department of the Environment. Remarkably, no attempt was made on the part of the Belgian organization to call on a partnership with the private sector to deal with the situation.

One senses from these recent developments that, from passive first and later reactive, the attitude of the Belgian authorities has now become overtly offensive. Changes in structure and leadership are taking place. The intention is to take the initiative whenever possible with all means available, in order to exploit to their full potential the powers the Administration has been granted at sea by law. The missing element is obviously a commensurate budgetary provision. The struggle for budgets has been fierce in recent years, and it could intensify in the future, as the lack of balance of the current situation becomes more obvious. Interestingly, this scanty policy may be encouraged by recent European legislation (the "Erika II package") that goes beyond the rationale of the International Maritime Organization (IMO) conventions and lays the full responsibility for remediation on the polluter. Assuming that all costs should be borne by the liable parties, the Government could feel it can afford to do with little initial fixed costs. This is contradictory of course, since the liability principle simply means that any reasonable cost incurred by the authorities, whether fixed or marginal, is fully reimbursable. That is, assuming the costs have been made in the first place. After all, the obligation of adequate preparedness for disaster management can only lie with the State. The new legislation at issue deserves some concluding remarks.

## 6.  The Future: Concluding Remarks

Two recent European Union (EU) directives have attracted much attention in the maritime world: Directive 2004/35/EC on environmental liability (EU, 2004) and Directive 2005/35/EC on ship-source pollution (EU, 2005). Directive 2004/35/EC states (art. 6; art. 8) that where environmental damage has occurred an operator shall, without delay, take the necessary remedial measures, and that he shall bear the costs for the preventive and remedial actions taken. Annex II of the directive states that the purpose of primary remediation is to restore the damaged natural resources to, or towards, baseline condition. Where primary remediation does not succeed, complementary and compensatory remediation must be undertaken to provide a similar level of natural resources, possibly at an alternative site, as well as additional improvements to protected habitats and species to compensate for the interim losses.

Although the directive does not apply to cases covered by the liability regime of the IMO conventions, it does apply to shipping when these conventions are not yet in force in the Member State, and to all cases not covered by the IMO conventions. The directive is viewed by marine experts as largely inapplicable to the marine environment because remediation measures going beyond cleanup are regarded as largely impractical except for selected shore habitats. Considerable concern has been expressed that Member States such as Belgium could be tempted to apply the directive indiscriminately simply to comply with the letter, disregarding the spirit of the text and neglecting the consequences.

The second directive, Directive 2005/35/EC, regards any ship-source discharge of polluting substances as infringement if committed with intent, recklessly or by serious negligence (art. 4). Such infringements are subject to penalties which may include criminal penalties (art. 8). The salvage industry has reacted with extreme worry and defiance to this text. It claims that recent events have shown that governments of Member States have not hesitated to detain the master of a vessel caught in a maritime accident simply because negligence on his part could not be excluded. If the Directive is implemented with its current wording, a similar course of action could be taken against a salvage master who has not succeeded in preventing pollution damage when dealing with a casualty. This is totally illogical and unacceptable.

It would in effect be the end of the salvage industry that simply could not afford such a large risk for what represents its specific business. A shipping industry coalition comprising the International Salvage Union (ISU), Intertanko, Intercargo, the Greek Shipping Cooperation Committee, and Lloyds Register filed an application for Judicial Review in the Administrative Court of the High Court of Justice in London on December 23, 2005. On June 30, the High Court ruled in favor of the coalition and granted a reference to the European Court of Justice for a determination on the legality of the Directive (URS, 2006). Those proceedings are still pending.

Thus, it seems that the general offensive rightly launched by the European and Belgian authorities against accidental marine pollution and its ecological consequences has now reached a point of contention. Having succeeded in convincing the shipping industry of the need to prevent and remedy marine pollution, and having done much to obtain its co-operation on joint schemes of action, governments could now be tempted to be too demanding for what the industry can sustain. Having made the industry a (near) ally, would it not be a shame to now loose it again? Would endless litigations in the future serve the best interests of the State and the environment? Is the current strategy right?

The national strategy in Belgium is bound to evolve further as the country's Coast Guard arrangements shape up and gain substance in the future. Hopefully, this will include some understanding with the private sector for mutually advantageous partnerships, adequate budgetary resources to anticipate the needs of counter-pollution intervenetions, and vessels that can move effectively into a spill to deploy the combating equipment and mitigate environmental damage.

## References

Anatec, 2003, *Shipping Traffic – Thornton Bank*. January (cited by Germanischer Lloyd Offshore and Industrial Services, 2003).

Bonn Agreement, 1983, Agreement for co-operation in dealing with pollution of the North Sea by oil and other harmful substances. http://www.bonnagreement.org

Department of Transport, 1987, *mv Herald of Free Enterprise*. Report of Court No. 8074 Formal Investigation. Her Majesty's Stationary Office, London.

EU, 2004, Directive 2004/35/CE of 21 April 2004 on environmental liability with regard to the prevention and remedying of environmental damage. *Official Journal of the European Union*, L 143/56–75.

EU, 2005, Directive 2005/35 on ship-source pollution and on the introduction of penalties for infringements. *Official Journal of the European Union*, L 255/11–21.

Germanischer Lloyd Offshore and Industrial Services, 2003, *Offshore Wind Energy Park Thornton Bank Technical Risk Analysis*. Report N° GL O-03-391 Rev. 1, Department Risk Management & SHE.

Haelters, J., Vigin, L., Stienen, E.W.M., Scory, S., Kuijken, E., and Jacques, T.G., 2004, Importance ornithologique des espaces marins de la Belgique. *Bulletin de l'Institut Royal des Sciences Naturelles de Belgique*, *74* Suppl.

ITOPF, 1994, *Study of Oil Spill Response Capabilities in Belgium*. Etude réalisée pour la Fédération Pétrolière de Belgique, Bruxelles, Belgique.

Jacques, T.G., 1985, Scientific Evaluations of an Incident at Sea Involving a Sunken Ship Carrying a Dangerous Cargo. In R. Van Grieken and R. Wollast (Eds.). *Progress in Belgian Oceanographic Research*, pp. 343–357. The University of Antwerp (UIA), Antwerp, Belgium.

Jacques, T.G., 1990, The *Herald of Free Enterprise* Accident: the Environmental Perspective. *Oil and Chemical Pollution*, 6:55–68.

Jacques, T.G., 1995, La pratique belge et le droit de la mer : la protection de l'environnement marin. *Revue Belge de Droit International, 1995/1*:127–146.

MMM (Marien Milieu Marin), 1999, Loi du 20 janvier 1999 visant la protection du milieu marin dans les espaces marins sous juridiction de la Belgique. *Moniteur Belge 50* (12 mars 1999), pp. 8033–8054.

Norcontrol IT, 2006, http://www.norcontrolit.com/newsrelease20020422.htm

UNCLOS, 1982, Loi du 18 juin 1998 portant assentiment à la Convention des Nations Unies sur le droit de la mer, faite à Montégo Bay le 10 décembre 1982 et l'Accord relatif à l'application de la partie XI de la Convention des Nations Unies sur le droit de la mer du 10 décembre 1982, fait à New York le 28 juillet 1994. *Moniteur Belge, 183* (16 septembre 1999), pp. 34484–34643.

URS, 2006, Union de Remorquage et de Sauvetage – Towage and Salvage Union, Antwerp, Belgium, personal communication.

**Copyright for illustrations:** Management Unit of the North Sea Mathematical Models (MUMM), Brussels.

# DIMENSIONING OF NORWEGIAN OIL SPILL PREPAREDNESS: FOCUSING ON THE ARCTIC NORTH NORWAY AND THE BARENTS SEA

J.M. LY[†]
*Norwegian Coastal Administration, Department of Emergency Response, PO Box 125, NO-3191 Horten, Norway*

**Abstract***. The Norwegian Coastal Administration (NCA) is the authority responsible for governmental oil spill response preparedness in Norway. The presentation will give a general introduction explaining how Norwegian oil spill preparedness is currently organized, and how the NCA use an environmental risk based approach to assess oil spill contingency. The presentation will briefly describe the methodology and approaches used to estimate and identify the specific preparedness level and amount of equipment and resources. Finally the operational preparedness, challenges and a description of the cooperation with Russia in the Barents Sea area is provided.

**Keywords:** oil spill response, arctic, Norway, preparedness

---

[†] To whom correspondence should be addressed. E-mail: johan-marius.ly@kystverket.no

* Full presentation available in PDF format on CD insert.

W. F. Davidson, K. Lee and A. Cogswell (eds.), *Oil Spill Response: A Global Perspective.*   289
© Springer Science + Business Media B.V. 2008

# REGIONAL CITIZENS' ADVISORY COUNCILS: ENSURING SAFE TRANSPORT OF CRUDE OIL IN ALASKA

J.S. FRENCH[†] & L. ROBINSON
*Prince William Sound Regional Citizens' Advisory
Council, 3709 Spenard Rd. Suite 100, Anchorage, AK
99503, USA*

**Abstract.** The role of citizens' groups in advising industry and government regulators has increased steadily in the United States over the past two decades. Prior to the *T/V Exxon Valdez* oil spill (EVOS) in 1989, various citizens groups expressed increasing concern about complacency and lax regulation by oil shippers and regulators but their attempts to form citizen advisory groups were unsuccessful. On March 29, 1989, *T/V Exxon Valdez* ran aground on Bligh Reef spilling 40 million liters of crude oil into the waters of Prince William Sound. The incident changed forever how crude oil is transported in the United States. Among other changes, The Oil Pollution Act (OPA) of 1990 established two industry-funded Regional Citizens' Advisory Councils (RCACs). One is designated for Prince William Sound (PWS) and the downstream regions affected by EVOS. The other was for Cook Inlet (CI) and downstream communities. Both RCACs are non-profit corporations with Boards of Directors representing both the governmental and other stakeholder entities within their region, and the various regulatory entities as ex-officio members. The member entities of both RCACs include tourism, recreational user, and environmental groups as well as commercial fishing and aquaculture interests. The governmental entities in CIRCAC represent a much larger, more urban population than those in PWSRCAC. Also four board seats in PWSRCAC represent primarily Alaska Native interests while CIRCAC has one. Cook Inlet and PWS are very different bodies of water which present different challenges to the safe transport of crude oil. The oil producing and transportation facilities within the areas of responsibility of the two RCACs are also quite different. CIRCAC is concerned with offshore production as well as transportation by multiple funding entities.

---

[†] To whom correspondence should be addressed. E-mail: jsfrench@arctic.net

W. F. Davidson, K. Lee and A. Cogswell (eds.), *Oil Spill Response: A Global Perspective.* 291
© Springer Science + Business Media B.V. 2008

PWSRCAC is better funded with more focused responsibilities. The two RCACs collaborate on projects of mutual interest, such as geographical response strategies, ports of refuge, and shore zone mapping. They have chosen different paths regarding environmental monitoring, issues, and regulations more specific to their regions of interest.

**Keywords:** citizen's advisory councils, transport, Exxon Valdez, oil spill response

# THE ROLE OF THE REGIONAL ENVIRONMENTAL EMERGENCIES TEAM (REET) IN EMERGENCY RESPONSE IN THE ATLANTIC REGION OF CANADA

R. PERCY[†] & S. DEWIS

*Environment Canada, 16th Floor Queen Square,*
*Dartmouth, Nova Scotia, B2Y 2N6, Canada*

**Abstract**[*]. A significant risk of a marine oil spill exists along the east coast of Canada due to increasing ship traffic and offshore oil exploration and development. Combine this with the natural hazards (such as weather and ice) associated with the North Atlantic Ocean and the presence of significant natural resources (such as wildlife and fishery) and the potential for a major oil spill with significant environmental impacts is created. Based on the lessons learned from past spills, the oil spill response community in the Atlantic Region has evolved an oil spill response network that focuses on improving the various procedures relating to spill prevention, preparedness, response, and damage restoration. An active player in these activities is the Regional Environmental Emergencies Team or REET. This paper will provided a brief overview of the role of the "REET" in emergency response in the region. Several of the key aspects of the REET approach will be described including the partnerships which are created, the technical and scientific information which is available from various organizations and the value of providing coordinated and consolidated environmental advice to the OSC. Several short case studies will be presented to describe the recent activities of the REET.

**Keywords:** REET, oil spill response, prevention

[†] To whom correspondence should be addressed. E-mail: roger.percy@ec.gc.ca
[*] Full presentation available in PDF format on CD insert.

# MINIMIZING ENVIRONMENTAL IMPACT OF OIL SPILLS: STATOIL'S R&D POSITION AND PRIORITIES

H.G. JOHNSEN[†] & J.E. VINDSTAD
*Senior research scientist, Environmental Technology,*
*Statoil Research Centre, Arkitekt Ebbellsvei 10, Rotvoll,*
*N-7005 Trondheim, Norway*

**Abstract.** Statoil ASA has during the last decade carried out a specific research program, Arctic Technology, aimed at developing techno-logies for offshore areas with cold climate and ice. For the development and operations of any Arctic hydrocarbon field, both on the Norwegian Continental Shelf (Lofoten and Barents Sea) and internationally (e.g., Russian Barents Sea, North Caspian Sea, Sakhalin), safety and envi-ronmental sustainability are key aspects. Environmental risk assessment and management of oil spill contingency and response is for this reason is the focus of Statoil's Arctic Technology project. Statoil has identified three main directions of research within this area: improve environ-mental risk assessment tools and underlying models, improve instal-lations with respect to accidental probability, and improve oil spill response technologies for cold and ice-covered areas. For assessing the risk related to accidental discharges the Environmental Impact Factor Model for risk assessment of acute oil spills (EIF Acute) has recently been developed. The EIF Acute Model is based on the current practice of environmental risk assessment on the Norwegian sector (MIRA), with several improvements, especially regarding the tools applicability to serve as a management tool, where risk reducing efforts (both reducing probability and the consequence) could be taken into account. Statoil is also involved in several joint industry technology projects with the objective to improve tools and technologies for oil spill contingency and response in Arctic areas. Research has been focused on the following technologies: dispersant use in cold and arctic environment; expanding the window for *in-situ* burning; and, deve-lopment of technologies for remote sensing and surveillance of oil in and under ice. Statoil regards our involvement in these activities as a

---

[†] To whom correspondence should be addressed. E-mail: hanjo@statoil.com

W. F. Davidson, K. Lee and A. Cogswell (eds.), *Oil Spill Response: A Global Perspective.*   295
© Springer Science + Business Media B.V. 2008

way to meet environmental challenges and prepare the company for activities as an operator in the arctic environment. The activities will increase competence in management of acute discharges within cold and ice-covered areas and will lead to improved company standards and performance within the Health, Safety and Environment (HSE) areas. This development is also strongly related to the "zero harm mindset" as the overall foundation of the company's HSE goals.

**Keywords:** arctic, environmental risk assessment, oil spill response

# MODELING IMPACTS AND TRADEOFFS OF DISPERSANT USE ON OIL SPILLS

D. FRENCH-MCCAY[†]
*J.J. Rowe Applied Science Associates, Inc., 70 Dean Knauss Drive, Narragansett, 02882-1143, Rhode Island, USA*

W. NORDHAUSEN
*Office of Spill Prevention and Response, San Diego, California, USA*

J.R. PAYNE
*Payne Environmental Consultants, Inc.,1991 Village Park Way, Suite 206 B, Encinitas, CA 92024, USA*

**Abstract.** Successful application of dispersants can reduce oil-spill impacts to wildlife and shoreline habitats, with the tradeoff of dispersed oil potentially causing impacts to water column organisms. Oil-spill fate and transport modeling was used to evaluate the maximum potential water column hydrocarbon concentrations and impacts of oil spills with dispersant use in offshore waters. The model estimated expected concentrations in the surface mixed layer for the largest potential volume of oil that could be dispersed in U.S. waters at any one location and time, that dispersed by a single sortie of a C-130 aircraft assuming a 20:1 oil to dispersant ratio and 80% efficiency (378.5 $m^3$ of light Arabian crude oil). For this oil volume and no dispersant, wildlife impacts would occur on the scale of 100 s $km^2$, whereas upper (<20 m) water column effects with 80% dispersion in one contiguous area would occur on the scale of 1 $km^2$. The model results for these offshore scenarios show that the tradeoff of decreasing wildlife impacts with dispersant use at the expense of possibly increasing water column impacts is supportive of dispersant use. The exceptions would be when sensitive water column biota are present in surface water under the slick and in shallow (<10 m) confined water bodies where dilution would be slower.

---

[†] To whom correspondence should be addressed. E-mail: dfrench@appsci.com

W. F. Davidson, K. Lee and A. Cogswell (eds.), *Oil Spill Response: A Global Perspective.* 297
© Springer Science + Business Media B.V. 2008

**Keywords:** modeling, dispersants, oil spill response

## 1.  Introduction

To date, few oil spills in U.S. waters have been treated with chemical dispersants. However, the U.S. Coast Guard (USCG) is presently developing new regulations regarding response plan oil removal capacity (Caps) requirements for tank vessels and marine transportation-related facilities (based on Caps review; USCG, 1999), which are expected to increase use of dispersants. To support the new regulations, the USCG has recently issued a Draft Programmatic Environmental Impact Statement (PEIS; USCG, 2004). As part of that effort, modeling was used to evaluate the tradeoffs of potential increased impacts to water column biota while reducing impacts to wildlife and shoreline habitats (French McCay *et al.*, 2004). The conclusions of the modeling, as well as in recent analyses by others, were that there would be a net environmental benefit of using dispersants at least in offshore waters.

The application of dispersants may reduce impacts to wildlife (e.g., seabirds, sea otters) and shoreline habitats, but with the tradeoff that the dispersed oil may cause impacts to water column organisms. Aquatic organisms may be adversely impacted either directly or via the food web by the toxic effects of oil components that enter the water column, particularly the soluble compounds (i.e., monoaromatic hydrocarbons, MAHs, and two-three ring polycyclic aromatic hydrocarbons, PAHs). Evaluations of bioassays and modeling have shown that while the MAHs are dissolved in higher concentrations into water, the PAHs (and particularly the alkyl-substituted homologues) are more toxic and may affect biota via dissolved concentrations or by uptake from dispersed oil droplets (French McCay, 2002, 2003). Other soluble and semi-soluble hydrocarbons in oil may also contribute to aquatic toxicity (NRC, 2005). Non-aromatic hydrocarbons are much less soluble, so are not bioavailable.

Concentrations of oil hydrocarbons in water are a complex function of environmental conditions (e.g., wind, turbulence, temperature) and dilution volume (volume of water into which the oil is dispersed). Overall, under natural conditions, adverse impacts increase with spill size. Nonetheless, there is great variability related to the environmental conditions after the spill. Aquatic organisms suffer more adverse impact under windy conditions where high waves mix unweathered oil into the water than in calm weather where little or no significant natural dispersion into the water column occurs (French McCay and Payne, 2001; French McCay, 2002, 2003). Dispersants lower the oil-water interfacial tension, which promotes increased entrainment and dissolution of oil components into the water column. Use of dispersants on fresh oil under light wind conditions could potentially increase the water column impact analogous to those under windy conditions where natural dispersion occurs, while weathering before dispersants are applied reduces the concentrations of MAHs and PAHs in the surface oil and consequently in the water column (French McCay and Payne, 2001).

The oil spill modeling was performed using SIMAP (French McCay, 2003, 2004), utilizing wind data, current data, and transport and weathering algorithms to calculate the mass of oil components in various environmental compartments (water surface, shoreline, water column, atmosphere, sediments, etc.), oil pathway over time (trajectory), surface oil distribution, and concentrations of the oil components in water and sediments. SIMAP was derived from the physical fates and biological effects submodels in the Natural Resource Damage Assessment Model for Coastal and Marine Environments (NRDAM/ CME), which were developed for the U.S. Department of the Interior (USDOI) as the basis of Comprehensive Environmental Response, Compensation and Liability Act of 1980 (CERCLA) Natural Resource Damage Assessment (NRDA) regulations for Type A assessments (French *et al.*, 1996). SIMAP has been further developed for use in impact and ecological risk assessments, cost-benefit analyses, and spill-response planning, as well as natural resource damage assessment.

## 2.  Model Description

SIMAP contains physical fate and biological effects models, which estimate exposure and impact on each habitat and species (or species group) in the area of the spill. Technical documentation is in French McCay (2003, 2004).

### 2.1. PHYSICAL FATES MODEL

The physical fate model estimates the distribution of oil (as mass and concentrations) on the water surface, on shorelines, in the water column, and in the sediments. Processes simulated include slick spreading, evaporation of volatiles from surface oil, transport on the water surface and in the water column, randomized dispersion, emulsification, entrainment of oil as droplets into the water column, resurfacing of larger droplets, dissolution of soluble components, volatilization from the water column, partitioning, sedimentation, stranding on shorelines, and degradation. Oil mass is tracked separately for lower-molecular-weight aromatics (1–3-ring aromatics), which are soluble and cause toxicity to aquatic organisms (French McCay, 2002), other volatiles, and non-volatiles. Lower molecular weight aromatics dissolve both from the surface oil slick and whole oil droplets in the water column, and are partitioned in the water column and sediments according to equilibrium partitioning theory (French *et al.*, 1996; French McCay, 2003, 2004).

"Whole" oil (containing non-volatiles and volatile components not yet volatilized or dissolved from the oil) is simulated as floating slicks, emulsions and/or tarballs, or as dispersed oil droplets of varying diameter (some of which may resurface). Sublots of the spilled oil are represented by Lagrangian elements ("spillets"), each characterized by mass of hydrocarbon components and water content, location, thickness, diameter, density, and viscosity. Spreading (gravitational and by transport processes), emulsification, weathering (volatilization and dissolution loss), entrainment, resurfacing, and transport processes determine the thickness, dimensions, and locations of floating oil over time. Output of the fate model includes the location, dimensions, and physical-chemical characteristics over time of each spillet representing the oil (French McCay, 2003, 2004).

Concentrations in the water column are calculated by summing mass (in Lagrangian particles) within each grid cell of a 100 (east-west) by 100 (north-south) by 5 vertical layer grid scaled each time step to just cover the dimensions of the plume. This maximizes the resolution of the contour map at each time step and reduces error caused by averaging mass over large cell volumes. Distribution of mass around the particle center is described as Gaussian in three dimensions, with one standard deviation equal to twice the diffusive distance ($2D_x t$ in the horizontal and $2D_z t$ in the vertical, where $D_x$ is the horizontal and $D_z$ is the vertical diffusion coefficient, and t is particle age). The plume grid edges are set at one standard deviation out from the outer-most particle. Concentrations of particulate (oil droplet) and dissolved aromatic concentrations are calculated in each cell and time step and saved to files for later viewing and calculations. These data are used by the biological effects model to evaluate exposure, toxicity, and effects.

The physical fates model has been validated with more than 20 case histories, including the *Exxon Valdez* and other large spills (French McCay, 2003, 2004; French McCay and Rowe, 2004), as well as test spills designed to verify the model's transport algorithms (French *et al.*, 1997).

## 2.2. BIOLOGICAL EFFECTS MODEL

The biological exposure model in SIMAP estimates the area, volume, or portion of a population affected by surface oil, concentrations of oil components in the water, and sediment contamination (French McCay, 2003, 2004). For wildlife (birds, mammals, and sea turtles), the number or fraction of a population suffering oil-induced effects is proportional to the water-surface area swept by oil of sufficient quantity to provide a lethal or sublethal dose to an exposed animal. The probability of exposure is related to behavior: i.e., the habitats used and percentage of the time spent in those habitats on the surface of the water. Thus, an exposure index for seabirds and other offshore wildlife is the water area swept by more than 10-μm thick ($>10$ $g/m^2$) oil (which is sufficient to provide a lethal dose, French *et al.*, 1996; French McCay and Rowe, 2004). For shorebirds and other wildlife on or along the shore, an exposure index is length of shoreline oiled by $>10$ $g/m^2$.

The most toxic components of oil to water column and benthic organisms are low molecular weight compounds, which are both volatile and soluble in water, especially the aromatic compounds (NRC, 1985, 2002; French McCay, 2002). Organisms must be exposed to hydrocarbons in order for uptake to occur and aquatic biota are exposed primarily to hydrocarbons (primarily aromatics) dissolved in water. Thus, exposure and potential effects to water column and bottom-dwelling aquatic organisms are related to concentrations of dissolved aromatics in the water. Theoretically, exposure to microscopic oil droplets could also impact aquatic biota either mechanically (especially filter feeders) or as a conduit for exposure to semi-soluble aromatics (taken up via the gills or digestive tract). The effects of the dissolved hydrocarbon components are additive.

Mortality is a function of duration of exposure – the longer the duration of exposure, the lower the effects concentration (see review in French McCay, 2002). At a given concentration after a certain period of time, all individuals that will die have done so. The LC50 is the lethal concentration to 50% of exposed organisms. The incipient LC50 ($LC50_\infty$) is the asymptotic LC50 reached after infinite exposure time (or long enough that that level is approached). Percent mortality is a log-normal function of concentration, with the LC50 the center of the distribution.

The value of $LC50_\infty$ ranges from 5 to 400 µg/L for 95% of species exposed to dissolved PAH mixtures for over 96 h (French McCay, 2002). The $LC50_\infty$ for the average species is about 50 µg/L of dissolved PAH. These LC50 values have been validated with oil bioassay data (French McCay, 2002), as well as in an application of SIMAP to the *North Cape* oil spill where field and model estimates of lobster impacts were within 10% of each other (French McCay, 2003).

In SIMAP, aquatic organisms are modeled using Lagrangian particles representing schools or groups of individuals. Pre-spill densities of fish, invertebrates, and wildlife (birds, mammals, reptiles, and amphibians) are assumed evenly distributed across each habitat type defined in the application of the model. (Habitat types may be defined to resolve areas of differing density for each species, and the impact in each habitat type is then separately computed.) Mobile fish, invertebrates, and wildlife are assumed to move at random within each habitat during the simulation period. Benthic organisms either move or remain stationary on/in the bottom. Planktonic stages, such as pelagic fish eggs, larvae, and juveniles (i.e., young-of-the-year during their pelagic stage(s)), move with the currents.

Mortality of fish, invertebrates, and their eggs and larvae was comuted as a function of temperature, concentration, and time of exposure. Percent mortality was estimated for each of a large number of Lagangian particles representing organisms of a particular behavior class (i.e., planktonic, demersal, and benthic, or fish classed as small pelagic, large pelagic, or demersal). For each Lagrangian particle, the model evaluates exposure duration, and corrects the LC50 for exposure time and temperature (French McCay, 2002) to calculate mortality. The percent mortalities were summed, weighed by the area represented by each Lagrangian particle to estimate a total equivalent volume for

100% mortality. In this way, mortality was estimated on a volume basis, rather than necessitating estimates of species densities to evaluate potential impacts. In addition to mortality estimates, the volume exceeding 1 μg/L total dissolved aromatics was used as an index for exposure for fish, invertebrates, and plankton. The algorithms for these calculations and their validation are described in French McCay (2002, 2003, 2004).

## 3.  Model Inputs and Scenarios

### 3.1.  MATRIX OF RUNS

Below is an outline of the inputs that were varied in the model runs, with the basis of the assumptions described in following sections. The objective was to estimate the *maximum* possible contamination in the water column in any one general location that could occur if dispersants were applied, and compare that to the same scenario without dispersant application.

- One representative location off the coast of southern California: The objective is to estimate concentrations in space and time, water volumes exceeding thresholds of concern, and potential water column effects (as volume equivalents where toxicity would be expected). The results would be similar in most offshore areas of similar environmental conditions.

- One oil type: Light Arabian crude (a common crude oil transported in California waters; does not emulsify enough to prevent diserant use).

- One spill volume: maximum volume of oil that could be dispersed by a single sortie of a C-130 (378.5 $m^3$ [100,000 gal] of oil dispersed at 80%, 45%, or 20% efficiency).

- One oil thickness: median value for dispersant application (100 μm).

- No-dispersant use as compared to two potential times after oil was spilled when dispersant was applied: weathered 8 or 16 h to allow volatilization of many lower-molecular-weight components, but not so weathered that dispersants would be ineffective).

- Two wind speeds with corresponding natural diffusion rates (horiontal and vertical eddy diffusion coefficients): (1) 2.5 m/s (5 kts) with horizontal eddy diffusion 1 m$^2$/s, and (2) 7.5 m/s (15 kts) with horizontal eddy diffusion 10 m$^2$/s. Wind direction assumed constant from NNW so oil was transported southward along the coast. (Water plume dimensions and characteristics would be similar for all wind directions at the same angle relative to currents in offshore areas where dispersants might be applied, but the plume would move in different directions).

- Water depth in the surface mixed layer, assumed to retain all the dispersed oil (a worst-case condition): (1) 10 m deep (all cases), and (2) 20 m deep (selected cases).

- Background currents (other than surface wind drift from local winds) – constant in time and space: (1) 0 (none); (2) 13 cm/s (0.25 kt) to SSE (uniform in time and space), a typical current speed based on drifter studies during periods when the California Current prevails; and (3) 13 cm/s (0.25 kt) to NNW (uniform in time and space), a typical current speed based on drifter studies during reversal periods for the California Current.

- LC50s for acute toxic effects – range of species and life-stage sensitivities: (1) 2.5th percentile sensitivity – 5 µg/L (ppb) of PAH; (2) average sensitivity – 50 µg/L (ppb) of PAH, and (3) 97.5th percentile sensitivity – 400 µg/L (ppb) of PAH.

## 3.2. SPILL LOCATION AND CONDITIONS

For the spill site, a generic site offshore with a 10-m surface mixed layer was used. The oil is assumed not to diffuse below the surface mixed layer (to evaluate a worst-case condition of restricted vertical mixing). Additional runs were made assuming a 20-m surface mixed layer. Temperature is assumed 15°C and salinity 33 ppt, typical for California waters (French *et al.*, 1996). Air immediately above the water was assumed to have the same temperature as the water surface. Thus, water temperature was used in the model for calculation of evaporation rates of volatile and semi-volatile components.

Two wind speed conditions (constant over time) were run: (1) 2.5 m/s (5 kts), where natural entrainment is negligible, and (2) 7.5 m/s (15 kts) where natural wind-driven entrainment is significant. When winds were assumed 2.5 m/s, the horizontal eddy diffusion coefficient was 1.0 m²/s, a value typical of low turbulence conditions. When winds were assumed 7.5 m/s, the horizontal eddy diffusion coefficient was assumed 10 m²/s. The vertical eddy diffusion coefficient was calculated from wind speed in the wave-mixed layer (based on Thorpe, 1984; i.e., 1.5 times wave height, related to wind speed and duration) and assumed 1.0 cm²/s in deeper water. Using these assumptions, the dispersed oil was mixed over the surface mixed layer in a few hours (and restricted from mixing deeper by the assumption noted above).

For the base runs, the only currents included in the simulations are the wind drift in the surface mixed layer (calculated based on Youssef and Spaulding, 1993, 1994). Background currents would carry the oil plume downstream at the current speed, but concentrations would remain similar to those simulated with no current unless current shear disperses (dilutes) the plume faster than the wind mixing would alone. Additional runs were made assuming typical current speeds of 13 cm/s (0.25 kt) directed either with or against the wind.

### 3.3.  OIL VOLUME, WEATHERING TIMES, AND DISPERSANT APPLICATION ASSUMPTIONS

The payload volume for a typical C-130 ADDS pack is about 18.9 m³ (5,000 gal) and the application rate is 0.38–3.03 m³ (100–800 gal/min; S.L. Ross Environmental Research, 1997; Al Allen, September 2004, personal communication). Assuming 20:1 oil: dispersant ratio and 80% efficiency (USCG, 1999), the maximum amount of oil that could be dispersed in one location is 302.8 m³ (80,000 gal = 1,905 bbl = 261.1 MT) in 6–50 min. The dispersed oil simulations were made starting with 378.5 m³ (100,000 gal = 2,381 bbl = 326.3 MT) of oil released instantaneously and 80% of it chemically dispersed immediately, with the application finishing by 0.5 h. Model runs were previously performed to calculate the weathering that would occur by the time of the assumed dispersion, i.e., at 8 or 16 h. In order to provide comparison data, parallel model runs to the dispersed oil simulations were made assuming no dispersant applied. Time "zero" in the simulations (time of dispersant application or initiation of the no-dispersant control runs) was after 8 or 16 h of surface-oil weathering under the described conditions.

## 3.4. OIL PROPERTIES

Properties of Arabian Light crude oil were based on data in Environment Canada's Oil Property Catalogue (described in Jokuty *et al.*, 1999; data obtained September 2004 from http://www.etcentre.rg/ spills). The key properties at the standard temperature of 25°C were: density = 0.8641 g/cm$^3$, viscosity = 13 cp, oil-water surface tension = 21.6 dyne/cm and maximum mousse water content = 90%.

The properties of the unweathered and weathered oil (calculated using SIMAP, simulating 8 or 16 h of weathering at 15°C in 2.5 m/s wind; see below) are in Table 1. By 8 h after release, two thirds of the MAHs were evaporated, and by 16 h after release only 5% of the MAHs remained in floating oil. Thus, when the oil was dispersed in the simulations, most of the resulting dissolved hydrocarbon conentrations were PAHs.

TABLE 1. Oil properties for Arabian light crude oils used in the simulations.

| Property | Un-weathered | Weathered 8 h | Weathered 16 h |
|---|---|---|---|
| Fraction: monoaromatic hydrocarbons (MAHs) | 0.019571 | 0.006654 | 0.001058 |
| Fraction: 2-ring aromatics (2-ring PAHs) | 0.001572 | 0.001501 | 0.001390 |
| Fraction: 3-ring aromatics (3-ring PAHs) | 0.006230 | 0.006159 | 0.006048 |
| Fraction: Total 1–3 ring aromatics | 0.027373 | 0.014315 | 0.008495 |
| Fraction: Aliphatic volatiles[a]: boiling point < 180°C | 0.139429 | 0.047406 | 0.007534 |
| Fraction: Aliphatic volatiles[a]: boiling point 180–264°C | 0.167188 | 0.159688 | 0.147787 |
| Fraction: Aliphatic volatiles[a]: boiling point 264–380°C | 0.133810 | 0.132296 | 0.129893 |
| Fraction: Total Aliphatic Volatiles[a]: boiling point < 380°C | 0.440427 | 0.339390 | 0.285214 |

[a]Environment Canada's Oil Property Catalogue provided aromatic and total hydrocarbon data for volatile fractions of unweathered oil. The aromatic hydrocarbon fraction was subtracted from the total hydrocarbon fraction to obtain the aliphatic fraction of unweathered oil (Jokuty *et al.*, 1999).

## 3.5. OIL THICKNESS AND INITIAL DIMENSIONS OF THE OIL SLICK

The oil was assumed to be 100 μm thick at the time it was dispersed. This is half of the maximum limit for dispersant application of 0.2 mm, as described in the API 2001 guide (API *et al.*, 2001). The oil would not emulsify enough by the time it was dispersed to inhibit the dispersion efficacy assumed. Based on the weathering model runs, the oil would not exceed dispersible viscosity limits by 24 h after a spill.

With an initial volume of 377.6 m$^3$, and an oil thickness of 100 μm, the initial area of the slick was 3.776 km$^2$. As an initial condition, it was assumed this area was circular.

## 3.6. PROCEDURE FOR MODEL RUNS

Cases (with no dispersant applied) were run at each wind speed to determine weathering rates and the composition of the oil at 8 and 16 h after release. The spill simulations with dispersants (as well as parallel no-dispersant cases) were started using oil initialized at the specified thickness of 100 μm, with the post-spreading minimum thickness set at 100 μm (so it no longer spreads), and pre-weathered as much as it would by time dispersed (8 or 16 h). The percentage of each of the volatile and semi-volatile components left at the time dispersed (8 or 16 h) was used to determine percentage in the oil to start the with-dispersant simulation. Oil was dispersed at 0–0.5 h (so the simulation started at the time dispersed.)

The volume released in each of the non-dispersed pre-weathering cases needed to be the right amount to have 377.6 m$^3$ of floating oil remaining by time of "dispersant" application. Several non-dispersed oil runs were made to quantify the weathering by 8 or 16 h. We ran these using a constant 5 kt wind. If winds were ≥15 kts, the oil would be naturally entrained and scattered, and the dispersant application would not be as concentrated as on a contiguous area with 377.6 m$^3$ of floating oil. With different spill volumes, the percentage evaporated is slightly different. So, the spill size and percentage weathered data for each of the 8 and 16 h dispersant application times were computed. We entered data for pre-weathered oil into the SIMAP oil database for each time dispersant could be applied.

Paired runs were made, one each with and without dispersant added. The release was 326.3 MT of oil at the water surface as an instantaneous spill. Dispersant was applied within hours 0–0.5 at a rate of 37.76 m³/h (522.2 MT/h) to disperse 80% of the oil. Alternate runs assuming 45% (293.74 MT/h) and 20% (130.55 MT/h) of the oil disersed (lower efficiencies) were made for comparison. The approriate oil for the weathering time and thickness was used in the simulations.

## 4. Results

### 4.1. OIL FATE AND PLUME DIMENSIONS OVER TIME

The mass balance of oil over time for the model scenarios run (Table 2) indicated most of the (remaining) volatilization occurred during the first 24 h after time "zero," the time dispersant was applied (8 or 16 h after release). In cases with no dispersant added, most of the mass was either floating or evaporated. For cases with dispersant added, about 80% was in the water column initially, and most of the rest was in the atmosphere or remained floating. Mass was slowly lost to "decay" (photo- and biodegradation, at rates as per French McCay, 2004).

Model results are summarized as volumes, areas, and dimensions of the dissolved aromatic plume > 1 ppb over time. Table 3 summarizes the maximum dimension at any time after the oil is dispersed (or released in the no-dispersant cases). The change in plume dimensions over time for model simulations with 8 or 16 h of pre-weathering, assuming no-current or a 0.25 kt current, restricting vertical diffusion within a 10- or 20-m mixed layer depth, in 2.5 m/s or 7.5 m/s winds, and with and without dispersant were calculated (not shown here).

With the same wind speed and dispersant condition, the cases have similar patterns. With a 2.5 m/s wind and no dispersant, the conentration plume is relatively small (Table 3) and short-lived (lasting a few hours and patchy in space and time). With a 2.5 m/s wind and dispersant applied, the concentration plume is sizable and concentrations remain above 1 ppb for greater than the 10-day model run (Table 3). In 7.5 m/s winds, natural dispersion is considerable, and addition of dispersant at 80% efficiency increases the volume affected by >1 ppb by a factor of 2–3. In the 7.5 m/s, no-dispersant cases, concentrations in the plume remain above 1 ppb for 2–3 days (Table 3). As in light-wind conditions, in the 7.5 m/s scenarios the duration of water column exposure to concentrations >1 ppb is significantly increased with dispersant use, but the plume (>1 ppb concentration) is dispersed by 3–6 days (Table 3). Dispersant application at lower efficiencies results in a proportionately smaller plume volume and shorter duration of exposure for water column biota (Table 3).

The results indicate that differences are subtle with degree of weathering over the range 8–16 h. The plume is generally smaller with more pre-weathering before dispersant application; however, in the high wind (7.5 m/s) cases, the more viscous weathered oil has also spread more before entrainment, creating a larger profile (area) for the plume.

For the spills without dispersant, increasing the mixed layer depth lessens the volume affected by >1 ppb because of faster dilution (Table 3); although in the 7.5 m/s cases, concentrations remain >1 ppb for 40–45 h with a 20-m mixed depth as opposed to 30–32 hs with a 10-m mixed depth (Table 3). For the spills with dispersant applied, increasing the mixed layer depth increases the volume affected by >1 ppb because it stretched (deepened) the plume without diluting it to below 1 ppb (Table 3). The 2.5 m/s wind, no-dispersant case after 8 h of weathering also shows this behavior (Table 3). This behavior would have different patterns using other thresholds, the higher the threshold the more easily dilution overcomes the stretching phenomenon. Background currents also stretch the plume (in the horizontal) in most, but not all cases. These effects are subtlety relative to other factors discussed above.

## 4.2. ACUTE TOXICITY AND MAXIMUM POTENTIAL IMPACTS

Table 4 summarizes the equivalent areas of 100% loss for both plankton (i.e., water column biota) and wildlife (primarily birds). The model runs were designed such that the slicks would remain offshore; thus, no shoreline impacts are predicted for these cases. The impact to plankton was calculated for a range of toxicity values characterizing 95% of species (French McCay, 2002). Percent loss in each affected volume was summed and divided by the mixed layer depth to calculate equivalent area of 100% mortality.

For wildlife, the areas in Table 4 are the water-surface area swept by oil of sufficient quantity to provide a lethal dose to an exposed animal. For each species exposed, these areas would be multiplied by the probability of exposure (based on behavior: i.e., the habitats used and percentage of the time spent in those habitats on the surface of the water), mortality rate once exposed (near 100% for birds and fur-bearers, lower for other wildlife), and animal density ($\#$/km$^2$) to esti-mate a loss in numbers killed (French McCay, 2003, 2004). Animal density is highly variable, and so comparisons are made here using equivalent areas of 100% loss for birds spending most of their time on the water surface. Losses for other species groups would be <100% in the impact area.

Plankton impact results are highly sensitive to the toxicity value assumed, and indicate much higher impact areas for sensitive species and stages than average or insensitive species. The potential impact on sensitive species of dispersant use in these *worst-case* application sce-narios is evident in the results. The impacted water volume for a sen-sitive (2.5th percentile) species is negligible in 2.5 m/s of wind with no dispersant, on the order of 1–2 million cubic meters in 7.5 m/s of wind for 378.5 m$^3$ (100,000 gal, 326.3 MT) of naturally-dispersed oil; 20–40 million cubic meters in 7.5 m/s of wind for 302.8 m$^3$ (80,000 gal, 261.0 MT) of chemically-dispersed oil (80% efficiency), and 70–200 million cubic meters in 2.5 m/s of wind for 302.8 m$^3$ of chemically-dispersed oil (80% efficiency). The impacted water volume for a species of average sensitivity (50th percentile) is negligible in all wind con-ditions with no dispersant use, on the order of 0.5–0.9 million cubic meters in 7.5 m/s of wind for 302.8 m$^3$ (80,000 gal, 261.0 MT) of chemically-dispersed oil (80% efficiency), and 6–20 million cubic meters in 2.5 m/s of wind for 302.8 m$^3$ of chemically-dispersed oil (80% efficiency). Thus, the highest impacts to plankton (and other water-column biota) are when chemical dispersant is applied under light wind conditions where dilution is relatively slow.

Impacts of floating oil to wildlife were over much larger areas than impacts to the water column of the surface mixed layer. For example, in 2.5 m/s winds, dispersant use at 80% efficiency would reduce the impact area for wildlife by a factor of 2.1–2.4 from about 200–210 km$^2$ to about 80–100 km$^2$ (assuming no background current), whereas the impacted area of the surface mixed layer of the water column with dispersant use at 80% efficiency is 0.5–2.0 km$^2$ for a species of average sensitivity (and up to 14 km$^2$ for sensitive species). In 7.5 m/s winds, dispersant use at 80% efficiency reduces the impact area for wildlife by a factor of 4–6 from about 390–425 $km^2$ to 70–110 $km^2$ (assuming no background current), while the water area impacted is 0.05–0.09 $km^2$ for a species of average sensitivity (and up to 2.2 $km^2$ for sensitive species). For the untreated cases with natural dispersion only, the area oiled is larger with higher winds because entrained oil resurfaces behind the leading edge of the slick and spreads the area of impact. Thus, the reduction of wildlife impact by use of dispersants is more dramatic in higher winds and with less impact on water-column biota.

The design of the modeling was to evaluate *worst-case* scenarios for water-column impacts. Thus, the results should not be considered typical of impacts that would occur if dispersants were applied to an oil spill. Also, it is important to note that while the volumes and areas of the water-column surface mixed layer described above appear very impressive, the areas affected for even the most sensitive species are generally less than 4 km on a side. Volumes and areas of water impacted would be much less if the oil was patchy or more spread out (because each patch would be a smaller volume and there would be more edge where mixing and dilution would occur). Nonetheless, the surface area swept by oil would not be much different if the oil were patchy, suggesting the wildlife impact area would be similar (although there may be differences in wildlife behavior in the two circumstances). Thus, the water-column impact volumes and areas are conservatively high in this analysis. Note again that the wildlife impact areas are for birds that spend 100% of the time on the water surface. For species that spend much of their time flying, roosting, or swimming under floating oil, impacts would be much less than 100% in the area affected.

Volumes of water impacted would also be less when the efficiency of the dispersant application was less than 80%. For example, in 2.5 m/s wind, the impacted water volume for a species of average sensitivity (for 302.8 $m^3$ of chemically-dispersed oil after being weathered 16 h, no currents, 10 m mixed depth) is 15 million cubic meters if efficiency is 80%, 8 million cubic meters if efficiency is 45%, and 1 million cubic meter if efficiency is 20%. For the same model scenarios except in 7.5 m/s wind, the impacted water volume for a species of average sensitivity is 780 thousand $m^3$ if efficiency is 80%, 200,000 $m^3$ if efficiency is 45%, and 4,000 $m^3$ if efficiency is 20%.

TABLE 2. Model scenarios run.

| Scenario | Wind (kt) | Hours of weathering | Horiz dispersion (m²/s) | Currents (dir. to) | Mixed layer depth (m) | Dispersed (0 = none; or 80, 45, 20%) |
|---|---|---|---|---|---|---|
| w8hr-5kt-h1-c0-10m-nd | 5 | 8 | 1 | 0 | 10 | 0 |
| w8hr-5kt-h1-c0-10m-wd | 5 | 8 | 1 | 0 | 10 | 80 |
| w8hr-5kt-h1-c0-20m-nd | 5 | 8 | 1 | 0 | 20 | 0 |
| w8hr-5kt-h1-c0-20m-wd | 5 | 8 | 1 | 0 | 20 | 80 |
| w8hr-15kt-h10-c0-10m-nd | 15 | 8 | 10 | 0 | 10 | 0 |
| w8hr-15kt-h10-c0-10m-wd | 15 | 8 | 10 | 0 | 10 | 80 |
| w8hr-15kt-h10-c0-20m-nd | 15 | 8 | 10 | 0 | 20 | 0 |
| w8hr-15kt-h10-c0-20m-wd | 15 | 8 | 10 | 0 | 10 | 80 |
| w16hr-5kt-h1-c0-10m-nd | 5 | 16 | 1 | 0 | 10 | 0 |
| w16hr-5kt-h1-c0-10m-wd | 5 | 16 | 1 | 0 | 10 | 80 |
| w16hr-5kt-h1-c0-20m-nd | 5 | 16 | 1 | 0 | 20 | 0 |
| w16hr-5kt-h1-c0-20m-wd | 5 | 16 | 1 | 0 | 20 | 80 |
| w16hr-5kt-h1-cnnw-10m-nd | 5 | 16 | 1 | NNW | 10 | 0 |
| w16hr-5kt-h1-cnnw-10m-wd | 5 | 16 | 1 | NNW | 10 | 80 |
| w16hr-5kt-h1-csse-10m-nd | 5 | 16 | 1 | SSE | 10 | 0 |
| w16hr-5kt-h1-csse-10m-wd | 5 | 16 | 1 | SSE | 10 | 80 |
| w16hr-15kt-h10-c0-10m-nd | 15 | 16 | 10 | 0 | 10 | 0 |

| | | | | | | |
|---|---|---|---|---|---|---|
| w16hr-15kt-h10-c0-10m-wd | 15 | 16 | 10 | 0 | 10 | 80 |
| w16hr-15kt-h10-c0-20m-nd | 15 | 16 | 10 | 0 | 20 | 0 |
| w16hr-15kt-h10-c0-20m-wd | 15 | 16 | 10 | 0 | 20 | 80 |
| w16hr-15kt-h10-cmnw-10m-nd | 15 | 16 | 10 | NNW | 10 | 0 |
| w16hr-15kt-h10-cmnw-10m-wd | 15 | 16 | 10 | NNW | 10 | 80 |
| w16hr-15kt-h10-csse-10m-nd | 15 | 16 | 10 | SSE | 10 | 0 |
| w16hr-15kt-h10-csse-10m-wd | 15 | 16 | 10 | SSE | 10 | 80 |
| w16hr-5kt-h1-c0-10m-wd45 | 5 | 16 | 1 | 0 | 10 | 45 |
| w16hr-5kt-h1-c0-10m-wd20 | 5 | 16 | 1 | 0 | 10 | 20 |
| w16hr-15kt-h10-c0-10m-wd45 | 15 | 16 | 10 | 0 | 10 | 45 |
| w16hr-15kt-h10-c0-10m-wd20 | 15 | 16 | 10 | 0 | 10 | 20 |

TABLE 3. Maximum dimensions of dissolved aromatic plume (>1 ppb) after weathered for indicated amount of time.

| Scenario | Maximum volume (thousands of m³) | Hours at maximum volume | Maximum area(thousands thousands of m²) | Hours at maximum area | Maximum length N-S (m) | Hours at maximum length N-S | Maximum length E-W (m) | Hours at maximum length E-W | Hours plume dispersed to <1 ppb |
|---|---|---|---|---|---|---|---|---|---|
| w8hr-5kt-h1-c0-10m-nd | 59 | 3.5 | 59 | 3.5 | 2,418 | 5 | 214 | 1.5 | 2.5 |
| w8hr-5kt-h1-c0-10m-wd | 294,942 | 82 | 41,926 | 134 | 41,082 | 238 | 3,284 | 84 | >240 |
| w8hr-5kt-h1-c0-20m-nd | 75 | 3.5 | 75 | 3.5 | 2,418 | 5 | 214 | 1.5 | 2.5 |
| w8hr-5kt-h1-c0-20m-wd | 350,730 | 206 | 37,187 | 76 | 38,671 | 220 | 4,239 | 154 | >240 |
| w8hr-15kt-h10-c0-10m-nd | 177,345 | 11.5 | 40,772 | 13.5 | 19,510 | 26 | 5,955 | 15 | 30 |
| w8hr-15kt-h10-c0-10m-wd | 559,883 | 37 | 80,964 | 38 | 52,651 | 96 | 5,763 | 19 | 106 |
| w8hr-15kt-h10-c0-20m-nd | 103,963 | 9 | 49,074 | 21 | 23,204 | 35 | 7,331 | 21 | 40 |
| w8hr-15kt-h10-c0-20m-wd | 559,883 | 37 | 80,964 | 38 | 52,651 | 96 | 5,763 | 19 | 106 |
| w16hr-5kt-h1-c0-10m-nd | 44 | 1 | 44 | 1 | 1,173 | 2 | 198 | 1.5 | 2.5 |
| w16hr-5kt-h1-c0-10m-wd | 296,712 | 86 | 45,956 | 90 | 38,332 | 240 | 4,156 | 172 | >240 |
| w16hr-5kt-h1-c0-20m-nd | 44 | 1 | 44 | 1 | 1,173 | 2 | 198 | 1.5 | 2.5 |
| w16hr-5kt-h1-c0-20m-wd | 413,799 | 240 | 38,011 | 86 | 42,076 | 238 | 2,981 | 130 | >240 |
| w16hr-5kt-h1-cnmw-10m-nd | 64 | 3 | 64 | 3 | 1,674 | 3.5 | 198 | 1.5 | 2.5 |
| w16hr-5kt-h1-cnmw-10m-wd | 256,132 | 74 | 37,984 | 92 | 34,231 | 96 | 3,495 | 110 | 136 |
| w16hr-5kt-h1-csse-10m-nd | 44 | 1 | 44 | 1 | 1,174 | 2 | 198 | 1.5 | 2.5 |
| w16hr-5kt-h1-csse-10m-wd | 312,721 | 172 | 46,843 | 170 | 71,394 | 240 | 3,199 | 98 | >240 |
| w16hr-15kt-h10-c0-10m-nd | 137,042 | 10 | 36,730 | 13.5 | 20,795 | 29 | 8,068 | 15.5 | 32 |
| w16hr-15kt-h10-c0-10m-wd | 471,411 | 23.5 | 68,225 | 40 | 46,029 | 100 | 6,390 | 29 | 110 |
| w16hr-15kt-h10-c0-20m-nd | 105,804 | 7.5 | 47,785 | 17 | 26,317 | 38 | 7,222 | 20 | 45 |

| | | | | | | | | | |
|---|---|---|---|---|---|---|---|---|---|
| w16hr-15kt-h10-c0-20m-wd | 539,407 | 26 | 108,089 | 38 | 51,438 | 112 | 8,024 | 45 | 126 |
| w16hr-15kt-h10-cnnw-10m-nd | 1,982,612 | 13 | 51,389 | 16 | 20,171 | 28 | 6,913 | 21.5 | 30 |
| w16hr-15kt-h10-cnnw-10m-wd | 550,149 | 24 | 81,385 | 34 | 55,654 | 80 | 6,246 | 52 | 82 |
| w16hr-15kt-h10-csse-10m-nd | 153,398 | 11.5 | 411,278 | 15.5 | 23,276 | 36 | 6,690 | 15.5 | 41 |
| w16hr-15kt-h10-csse-10m-wd | 491,388 | 34 | 84,308 | 44 | 42,572 | 134 | 8,185 | 42 | 90 |
| w16hr-5kt-h1-c0-10m-wd45 | 177,656 | 45 | 26,536 | 84 | 38,257 | 220 | 2,981 | 84 | >240 |
| w16hr-5kt-h1-c0-10m-wd20 | 89,010 | 34 | 15,347 | 34 | 36,494 | 110 | 2,428 | 52 | 134 |
| w16hr-15kt-h10-c0-10m-wd45 | 389,499 | 20 | 66,031 | 21 | 45,669 | 104 | 7,183 | 24 | 74 |
| w16hr-15kt-h10-c0-10m-wd20 | 253,287 | 15 | 68,660 | 21 | 31,185 | 52 | 8,035 | 46 | 58 |

TABLE 4. Equivalent area (km ) of 100% mortality.

| Scenario | Plankton - sensitive: LC50(inf) 5 ppb | Plankton-average: LC50(inf) 50 ppb | Plankton-insensitive: LC50(inf) 400 ppb | Wildlife – if 100% vulnerable[a] |
|---|---|---|---|---|
| w8hr-5kt-h1-c0-10m-nd | 0.000 | 0.000 | 0.000 | 198 |
| w8hr-5kt-h1-c0-10m-wd | 14.650 | 1.753 | 0.105 | 83 |
| w8hr-5kt-h1-c0-20m-nd | 0.000 | 0.000 | 0.000 | 197 |
| w8hr-5kt-h1-c0-20m-wd | 7.421 | 0.658 | 0.005 | 83 |
| w8hr-15kt-h10-c0-10m-nd | 0.188 | 0.000 | 0.000 | 425 |
| w8hr-15kt-h10-c0-10m-wd | 1.428 | 0.060 | 0.000 | 68 |
| w8hr-15kt-h10-c0-20m-nd | 0.031 | 0.000 | 0.000 | 406 |
| w8hr-15kt-h10-c0-20m-wd | 2.134 | 0.062 | 0.000 | 83 |
| w16hr-5kt-h1-c0-10m-nd | 0.000 | 0.000 | 0.000 | 209 |
| w16hr-5kt-h1-c0-10m-wd | 13.583 | 1.541 | 0.092 | 100 |
| w16hr-5kt-h1-c0-20m-nd | 0.000 | 0.000 | 0.000 | 204 |
| w16hr-5kt-h1-c0-20m-wd | 7.564 | 0.628 | 0.001 | 97 |
| w16hr-5kt-h1-cnnw-10m-nd | 0.000 | 0.000 | 0.000 | 248 |
| w16hr-5kt-h1-cnnw-10m-wd | 13.751 | 1.061 | 0.027 | 105 |
| w16hr-5kt-h1-csse-10m-nd | 0.000 | 0.000 | 0.000 | 373 |
| w16hr-5kt-h1-csse-10m-wd | 18.650 | 2.047 | 0.047 | 157 |
| w16hr-15kt-h10-c0-10m-nd | 0.169 | 0.000 | 0.000 | 391 |
| w16hr-15kt-h10-c0-10m-wd | 1.687 | 0.078 | 0.000 | 107 |
| w16hr-15kt-h10-c0-20m-nd | 0.027 | 0.000 | 0.000 | 408 |

| | | | |
|---|---|---|---|
| w16hr-15kt-h10-c0-20m-wd | 2.164 | 0.085 | 0.000 | 108 |
| w16hr-15kt-h10-cnnw-10m-nd | 0.127 | 0.000 | 0.000 | 524 |
| w16hr-15kt-h10-cnnw-10m-wd | 3.356 | 0.054 | 0.000 | 117 |
| w16hr-15kt-h10-csse-10m-nd | 0.122 | 0.000 | 0.000 | 318 |
| w16hr-15kt-h10-csse-10m-wd | 3.596 | 0.086 | 0.000 | 58 |
| w16hr-5kt-h1-c0-10m-wd45 | 9.368 | 0.779 | 0.010 | 158 |
| w16hr-5kt-h1-c0-10m-wd20 | 1.362 | 0.098 | 0.000 | 184 |
| w16hr-15kt-h10-c0-10m-wd45 | 1.128 | 0.020 | 0.000 | 196 |
| w16hr-15kt-h10-c0-10m-wd20 | 0.479 | 0.004 | 0.000 | 276 |

This area would be multiplied by probability of exposure, mortality rate once exposed, and density (#/km$^2$) to estimate a species impact.

## 5.  Discussion

### 5.1.  POTENTIAL IMPACTS OF DISPERSANT USE ON WATER COLUMN BIOTA

Significant water column impacts would be expected only after large crude (or light fuel) oil spills under certain conditions: surface releases under storm conditions (high turbulence), spills where dispersants are applied with high efficiency on large oil volumes, and subsurface releases (pipelines and blowouts). The larger the volume of PAH released (i.e., the combination of oil volume and PAH content) and the more turbulent the conditions, the more dissolution of the more toxic compounds into the water column. Thus, these conditions would lead to the highest exposure to the most toxic compounds.

Use of chemical dispersants on a large volume of oil concentrated in a relatively small area could lead to toxic concentrations in the surface mixed layer of the area where oil is entrained. However, in most (if not all) cases, the floating oil being dispersed will not be in a large contiguous area of the magnitude modeled here. Volumes of water where impacts would occur would be much less if the oil is patchy or more spread out, or in the cases where the efficiency of the dispersant application is <80%. The later conditions will be the norm when dispersants are applied in the field under less-than-perfect conditions, with imperfect knowledge of the location of the oil, and where oil has naturally broken up into patches and convergence zones.

The model results for these offshore scenarios showing the tradeoff of decreased wildlife impacts with dispersant use at the expense of possibly increasing water column impacts, expressed on an impacted-area basis, is very supportive of dispersant use. For the oil volume examined and assuming no dispersant use, wildlife impacts would occur on the scale of 100 s km$^2$, whereas water column effects with dispersant use and as a worst case would occur on the scale of 1 km$^2$ in the upper mixed layer (10–20 m deep). The exception to this support would be if sensitive water column biota are present in the area of the slick. Dilution would also be slower in confined water bodies than modeled here for offshore scenarios. Thus, the results and conclusions presented here apply to unconfined water bodies that are at least 10 m deep.

## 6. Acknowledgements

This research was supported by California Fish and Game, Office of Spill Prevention and Response (OSPR) contract P0375036. We thank Yvonne Addassi and Robin Lewis of OSPR for their valuable inputs to the study.

## References

Al Allen, 2004, Spilltec, Woodinville, WA, personal communication, September 2004.

American Petroleum Institute (API), National Oceanic and Atmospheric Administration (NOAA), U.S. Coast Guard (USCG), and U.S. Environmental Protection Agency (USEPA), 2001, *Characteristics of Response Strategies: A Guide for Spill Response Planning in Marine Environments.* Joint Publication of API, NOAA, USCG, and USEPA, June 2001.

French McCay, D. and Payne, J.R., 2001, Model of Oil Fate and Water Concentrations with and with out Application of Dispersants, in: *Proceedings of the 24th Arctic and Marine Oilspill Program (AMOP) Technical Seminar,* pp. 611–645. Emergencies Science Division, Environment Canada, Ottawa, ON,

French, D., Reed, M., Jayko, K., Feng, S., Rines, H., Pavignano, S., Isaji, T., Puckett, S., Keller, A., French III, F.W., Gifford, D., McCue, J., Brown, G., MacDonald, E., Quirk, J., Natzke, S., Bishop, R., Welsh, M., Phillips, M. and Ingram, B.S., 1996, *The CERCLA Type A Natural Resource Damage Assessment Model for Coastal and Marine Environments (NRDAM/CME), Technical Documentation, Vol. I–V.* Submitted to the Office of Environmental Policy and Compliance, U.S. Department of the Interior, Washington, DC.

French, D.P., Rines, H. and Masciangioli, P., 1997, Validation of an Orimulsion Spill Fates Model Using Observations from Field Test Spills, in: *Proceedings of the Twentieth Arctic and Marine Oilspill Program (AMOP) Technical Seminar*, pp. 933–961. Emergencies Science Division, Environment Canada, Ottawa, ON.

French McCay, D.P., 2002, Development and Application of an Oil Toxicity and Exposure Model, OilToxEx. *Environmental Toxicology and Chemistry.* **21**(10): 2080–2094.

French McCay, D.P., 2003, Development and Application of Damage Assessment Modeling: Example Assessment for the *North Cape* Oil Spill. *Marine Pollution Bulletin.* **47**(9–12):341–359.

French McCay, D.P., 2004, Oil Spill Impact Modeling: Development and Validation. *Environmental Toxicology and Chemistry.* **23**(10):2441–2456.

French McCay, D.P., Aurand, D., Michel, J., Unsworth, R., Whittier, N., Lord, C., Dalton, C., Levine, R., Rowe, J., Sankaranarayanan, S., Kim, H.-S., Piovesan, R. and Hitchings, M., 2004, *Oil Spills Fate and Effects Modeling for Alternative*

*Response Scenario.* Final Report to US Department of Transportation, Cambridge, MA and US Coast Guard, Washington, DC, submitted by Applied Science Associates, Narragansett, RI, USA, 6 volumes.

French McCay, D.P. and Rowe, J.J., 2004, Evaluation of Bird Impacts in Historical Oil Spill Cases Using the SIMAP Oil Spill Model, in: *Proceedings of the Twenty-Seventh Arctic and Marine Oil Spill Program (AMOP) Technical Seminar*, pp. 421–452. Emergencies Science Division, Environment Canada, Ottawa, ON.

Jokuty, P., Whiticar, S., Wang, Z., Fingas, M., Fieldhouse, B., Lambert, P. and Mullin, J., 1999, *Properties of Crude Oils and Oil Products.* Manuscript Report EE-165, Environmental Protection Service, Environment Canada, Ottawa, ON, (www.etcentre.org/spills).

National Research Council, 1985, *Oil in the Sea: Inputs, Fates and Effects.* National Academy Press, Washington, DC.

National Research Council, 2002, *Oil in the Sea III: Inputs, Fates and Effects.* National Academy Press, Washington, DC.

National Research Council (NRC), 2005, *Understanding Oil Spill Dispersants: Efficacy and Effects.* National Research Council, Ocean Studies Board, National Academies Press, Washington, DC.

S.L. Ross Environmental Research, 1997, *A Review of Dispersant Use on Spills of North Slope crude oil in Prince William Sound and the Gulf of Alaska.* Report prepared for Prince William Sound Regional Citizens' Advisory Council, Anchorage, Alaska.

Thorpe, S.A., 1984, On the Determination of Kv in the Near Surface Ocean from Acoustic Measurements of Bubbles. *American Meteorological Society*: 861–863.

U.S. Coast Guard (USCG), 1999, *Response Plan Equipment Caps Review: Are Changes to Current Mechanical Recovery, Dispersant, and in situ Burn Equipment Requirements Practicable?* United States Coast Guard, U.S. Department of Transportation, Washington, DC.

U.S. Coast Guard (USCG), 2004, *Draft Programmatic Environmental Impact Statement (PEIS) for Mechanical Recovery and Alternative Removal Technologies.* U.S. Department of Homeland Security, U.S. Coast Guard, Washington, DC.

Youssef, M. and Spaulding, M.L., 1993, Drift Current Under the Action of Wind Waves, in: *Proceedings of the Sixteenth Arctic and Marine Oilspill Program Technical Seminar*, pp. 587–615. Ottawa, ON, Canada.

Youssef, M. and Spaulding, M.L., 1994, Drift Current Under the Combined Action of Wind and Waves in Shallow Water, in: *Proceedings of the Seventeenth Arctic and Marine Oilspill Program (AMOP) Technical Seminar*, pp. 767–784. Ottawa, ON, Canada.

# DEVELOPMENT OF OPERATIONAL OCEAN FORECASTING SYSTEMS AND IMPACT ON OIL PLUME DRIFT CALCULATIONS

F.J.M. DAVIDSON
*Northwest Atlantic Fisheries Centre, St. John's,*
*Newfoundland, A1C 5X1, Canada*

A.W. RATSIMANDRESY
*Biol. and Phys. Oceanography Section, Fisheries and*
*Oceans Canada NAFC, East White, Hills Rd., P.O. Box*
*5667 St. John's, Newfoundland, A1C 5X1, Canada*

C. HANNAH[†]
*Bedford Institute of Oceanography, 1 Challenger Drive,*
*P.O. Box 1006, Dartmouth, Nova Scotia, B2Y 4A2,*
*Canada*

**Abstract.** We review current progress on an operational ocean forecasting system for the North West Atlantic. The Canadian Newfoundland Operational Ocean Forecasting System (C-NOOFS) is being developed under a national coordinated effort through the DFO Virtual Center for Ocean Modeling and Data Assimilation. The model development is reviewed along with the various data assimilation components and output. The model domain covers the North West Atlantic with open boundary conditions that are nested within a basin or global scale model from the French Operational Oceanography Service MERCATOR-Ocean. The model is forced by Canadian Meteorological Service wind forecasts with initial conditions taken from MERCATOR-Ocean ocean analysis. The initialization from a best estimate through ocean analysis with realistic boundary conditions as opposed to climatology will improve the model forecast. We demonstrate the improvement in oil drift dispersion by the use of ocean currents derived from

[†] To whom correspondence should be addressed. E-mail: HannahC@dfo-mpo.gc.ca

W. F. Davidson, K. Lee and A. Cogswell (eds.), *Oil Spill Response: A Global Perspective.* 321
© Springer Science + Business Media B.V. 2008

a data assimilative model that includes the assimilation of sea surface height from a Satellite Altimeter and *in-situ* data. Useful outputs from this operational ocean forecasting system include enhancing search and rescue drift calculations and oil spill dispersion in near real time. Furthermore, model hind casts permit the determination of potential oil drift envelopes for specific regions throughout the Newfoundland Shelf. This enables scenario testing to better prepare for environmental responses regarding oil spills.

**Keywords**: modeling, ocean forecasting, oil spill response

# EMERGENCY PREVENTION, PREPAREDNESS AND RESPONSE WORKING GROUP OF THE ARCTIC COUNCIL: RECENT MARINE ENVIRONMENTAL INITIATIVES IN THE ARCTIC COUNCIL

M. MEZA[†]
*United States Coast Guard\Office of Response Field Activities Directorate, 2100 2nd St. S.W, Washington, DC 20593-0001, USA*

A. TUCCI
*United States Coast Guard \ Office of Response Field Activities Directorate, United States Coast Guard\Commandant (G-MOR), 2100 2nd St. S.W, Washington, DC 20593-0001, USA*

**Abstract[*].** The Emergency Prevention, Preparedness and Response Working Group of the Arctic Council continues to function as the focus of Arctic Council work on preparedness and response to oil, hazsub, and radiological releases. In addition the Working Group has recently been assigned responsibility for disaster response. Among the current work projects just completed or underway are the following: Shoreline Cleanup and Assessment Manual – The manual has been completed under Canadian and US leadership and is being translated into Russian. This completes the last gap identified in the 1998 Environmental Risk Assessment.

- Oil Response Waste Disposal Guide – The guide is being developed under the leadership of Canada to address transboundary issues in waste disposal and mutual aid as well as available technologies for oily waste disposal.
- The Arctic Monitoring and Assessment Program (AMAP) of the Arctic Council has completed the final draft of the Oil and Gas Assessment. National assessments have been drafted with variable completeness.

---

[†] To whom correspondence should be addressed. E-mail: Mark.Meza@uscg.mil
[*] Full presentation available in PDF format on CD insert.

W. F. Davidson, K. Lee and A. Cogswell (eds.), *Oil Spill Response: A Global Perspective.*  323
© Springer Science + Business Media B.V. 2008

- Arctic Marine Shipping Assessment – The Assessment will address the growth of Arctic Marine Shipping and the attendant implications for environmental, development and response issues among others. The Assessment is Chaired by the Protection of the Marine Environment (PAME) Chair.

- Arctic Council circumpolar mapping project continues to evolve and address a Circumpolar Map of resources at risk under the leadership of Norway. Further work is being done in updating data sets, development of infrastructure and policy on data, and development of applications.

- Source Control Management Projects related to Radiological Releases – Source control projects in two phases have been completed with Phase I – Source Control Management and Prevention Strategies for Chlorine Handling – being involved with the Apatity Waterworks and Phase II – Risk Assessment at the NIIAR Fuel Research Department – being involved with the Dimitrovograd, Ulyanovsk Region.

- Radiological Related Training continues with ISO 14001 Training Programs Handling, with Risk Assessment Methodology and with a Table Top Emergency Exercise related to the transportation of radioactive material (26 July 2005) conducted by Arctic Military Environmental Cooperation (AMEC) Partners in Murmansk, Russian Federation.

- Community Information Initiatives Related to Radiological Releases – Outreach and information initiative continue with the Community Radiation Information Project in the Kola Area, led by the US and Russian Federation. Three brochures have been published on Industrial North Nuclear Technologies and Environment, Emergency Public Information, and the ABC of Radiation Protection.

- Arctic Rescue – Proposed by the Russian Federation, the concept of "Arctic Rescue" is designed to identify potential sites for a network of stations throughout the Arctic to provide emergency notification and potential response to incidents in the region. A workshop was held in Moscow to define international legal agreement already in place in relation to this concept and to determine a way forward with the concept if it is supportable.

- Expansion of EPPR Mandate – The EPPR Working Group mandate was expanded to include natural disasters such as forest fires. A survey of past natural disasters is being lead by Finland. In this regard, the Russian Federation also presented work on flood response and database tracking.

**Keywords:** oil spill response, arctic counsels, disaster response

# THE EASTERN MEDITERRANEAN-BLACK SEA SYSTEM WITH HIGH OIL SPILL RISK

T. ÇOKACAR[†]

*Tubitak Marmara Research Centre (mrc), Chemistry and Environment Institute, P.O. Box 21 41470, Gebze, Kocaeli, Turkey*

**Abstract.** The Eastern Mediterranean and Black Seas along with Bosphours and Dardanelles Straits are located along a heavy oil traffic route between Middle East/Russia and Western Europe/USA, and therefore are faced with a continuing danger of oil spill. As oil spills are directly connected to flow and stratification characteristics, the basic oceanographic features of the region relevant to oil spill problems are briefly reviewed. Attention is then drawn to the increasing trend of oil tanker traffic, important accidents in recent decades, and the position of Turkey to responsed oil spills is presented.

**Keywords:** oil spill response, shipping traffic

## 1. Introduction

Turkey is surrounded by seas with a heavy tanker traffic route (Figure 1). In the north, the Black Sea together with the Turkish Straits System and the Aegean Sea is the only way to transport the oil produced in Russia, Azerbaijan and Central Asia to the west. On the other hand, the Gulf of Iskenderun located at far end of the Northeastern Medi-terranean Sea has two major terminals for Caspian and Iraqi oils. In spite of high risks for oil spills, only limited research has been con-ducted thus far on the oil spill trend and risk analyses in the region, even though its catastrophic consequences in many critical regions, such as the City of Istanbul, has been anonymously admitted by autho-rities. Nonetheless, the region is relatively well-known in terms of

---

[†] To whom correspondence should be addressed. E-mail: Tulay.Cokacar@mam.gov.tr

W. F. Davidson, K. Lee and A. Cogswell (eds.), *Oil Spill Response: A Global Perspective.*   327
© Springer Science + Business Media B.V. 2008

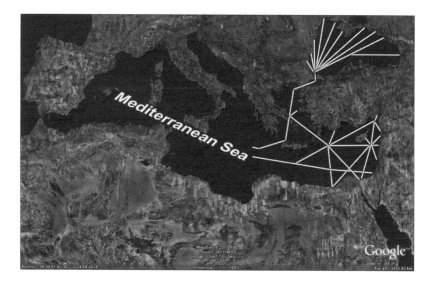

*Figure 1.* Oil tanker routes are drawn schematically according to LMIU 2001.
2: a review of regional characteristics.

its hydrodynamic characteristics, and there are some ongoing efforts
towards developing an operational near-real time prediction system.

The Eastern Mediterranean Basin and the Black Sea constitute two
largely isolated water bodies constrained by water exchanges through
straits. Both regions are highly sensitive to anthropogenic and climate-
induced variation, as the surrounding land mass is highly developed in
industry and tourism and they are in close proximity to major atmospheric centers of action (e.g., the North Atlantic Ocean, Sahara,
Indian Ocean). Their oceanographic features are briefly reviewed
below from an oil spill perspective.

## 1.1. BLACK SEA

With a surface area of 423,000 km$^2$, it is one of the world's largest
inland marine environments. Its connection to the Mediterranean Sea
is through two narrow straits; the Bosphorus and the Dardanelles. The
dissolved oxygen depletes at around 75–150 m depending on the region,
and the rest of the water column up to 2,000 m is anoxic. The Black

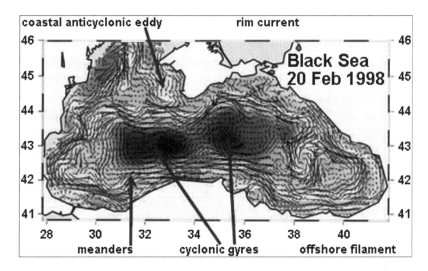

*Figure 2.* Modelled currents of Black Sea (Korotaev *et al.*, 2003)

Sea is known as the largest anoxic water body in the world. It has a positive water balance, with the excess of freshwater input form rivers and atmosphere through evaporation. The difference is balanced by the net outflow through the Bosphorus (Unluata *et al.*, 1990; Ozsoy and Unluata, 1997).

The fluxes through straits, sharp changes in topography, dynamic atmospheric forcing, and fresh water inputs from rivers are the principal factors governing the Black Sea circulation and thermohaline structure. The upper layer waters are characterized by a predominantly cyclonic, strongly time-dependent and spatially-structured basinwide circulation (Oguz *et al.*, 1994, 1998; Oguz and Besiktepe, 1999; Gawarkiewicz *et al.*, 1999; Krivosheya *et al.*, 2000). Both quasi-persistent and recurrent features of the circulation system are apparent in Figure 2. The Rim Current cyclonically encircles the basin flowing along the abruptly varying continental slope and margin topography. Two cyclonic sub-basin scale gyres comprising four or more gyres within the interior of the system and the a series of anticyclonic in the coastal side of the Rim Current zone are the characteristic features of the Black Sea circulation system. The complex, eddy-dominated circulation with different types of structural organizations within the interior cyclonic cell evolves continuously by interactions among its eddies as well as with meanders, and filaments of the Rim Current (Oguz *et al.*, 2005a). According to the Acoustic Doppler Current Profiler measurements (Oguz and Besiktepe, 1999), the Rim Current jet has a speed of 50–100 cm/s within the upper layer. In winter, the upper layer is homogenized up to 50 m depth with 5–6°C temperatures and 18.5–18.8 ppt salinities (Oguz *et al.*, 1990b; Krivosheya *et al.*, 2002). By the spring, as temperature in the surface mixed layer increases, the stratification begins to develop above a cold water lens, known as the Cold Intermediate Layer (CIL), being a remnant of the convectively generated cold water mass in winter. The summer mixed layer is typically shallower than 20 m depth, and characterized by temperatures up to 25°C and salinities around 18 ppt (Oguz *et al.*, 2005a).

The dominant wind direction is north and northeast in the western part, whereas southwesterlies dominate the eastern part of the basin. Gales from the northwest are common in winter (Ozsoy and Unluata, 1997). The air temperature has a strong north-south gradient in winter but is more uniform during summer (Ozsoy and Unluata, 1997). Air temperature decreases sharply in late October and November, and reaches a minimum of ~0°C in January and February (Ozsoy and Unluata, 1997).

## 1.2. TURKISH STRAITS SYSTEM

The Bosphorus and Dardanelles Straits together with the Marmara Sea (an area of 11,500 km$^2$) in between constitute the Turkish Straits System (TSS). The system possesses a two layer flow structure in which the lower layer flow is driven by the density differences between the Black Sea and the Aegean, and the upper layer flow due to higher sea level elevation of the Black Sea with respect to the Aegeran Sea (Ozsoy *et al.*, 1986; Oguz *et al.*, 1990a). The relatively fresh surface waters of Black Sea origin (average salinity of 18 ppt) flows through the Bosphorus into the Marmara Sea and Aegean Sea through the Dardanelles Straits. On the other hand, saline Mediterranean waters entering the Marmara Sea through the Dardanelles Strait flow in the opposite direction in the lower waters (average salinity of 35.5) through the Turkish Straits and eventually exits from the Bosphorus into the Black Sea. The flow in both Straits is hydraulically controlled. In the Bosphorus

Strait, three hydraulic controls exist due to contraction at the middle and the sills near the both ends, which develop the so-called maximal exchange flow conditions (Farmer and Armi, 1986; Armi and Farmer, 1987, Ozsoy et al., 1986; Unluata et al., 1990, Oguz, 2005b). The flow regime in the Dardanelles Strait differs from that of the Bosphorus by the presence of a single hydraulic control at the mid-strait constriction section, which implies a submaximal exchange (Ozsoy et al., 1986; Oguz and Sur, 1989). Due to its relatively wider and deeper morphology, frequent flow reversals occurs along both sides of the strait. Intense mixing between the layers takes place particularly in the vicinity of the hydraulic control (Ozsoy et al., 2002).

The exchange through the Straits determines the hydrography of the Marmara Sea. Black Sea waters entering from the Bosphorus with low salinity (18 ppt) above the salty Mediterranean waters (38 ppt) entering from the Dardanelles reveal a sharp density interface between the layers inhibiting mixing between the two layers. Nonetheless, considerable mixing occurs, especially in winter months due to cooling and wind-induced turbulent mixing across the layers, during which the upper layer salinity increases to $23 \pm 2$ ppt (Besiktepe et al., 1993). The entrainment process results in a lengthwise increase of upper layer salinity along the way to the Aegean exit (Unluata et al., 1990). The entrainment is particularly enhanced in the southern exit region of the Bosphorus where the upper layer flow enters into Marmara Sea in the form of a buoyant jet (Figure 3). The mean upper layer circulation in the Sea of Marmara is anti-cyclonic, mainly driven by the southward flowing Bosphorus jet in the enclosed domain. This anticyclonic gyre is modified by the Bosphorus jet during the high outflow conditions (spring and early summer) and by the wind forcing during the winter (Beşiktepe et al., 1994). A typical upper layer circulation structure for the early spring period deduced from the ADCP and CTD measurements is illustrated in Figure 3.

Northeasterly winds are prevalent throughout the year, with an average frequency of 60%, and southwesterly winds are of secondary importance (20%) (Beşiktepe et al., 1994). The daily average wind speed is 4 m s$^{-1}$. Strong wind events with typical speeds of 8–25 m s$^{-1}$ and durations of about 16 h occur in winter, especially near the Bosphorus junction (De Filippi et al., 1986).

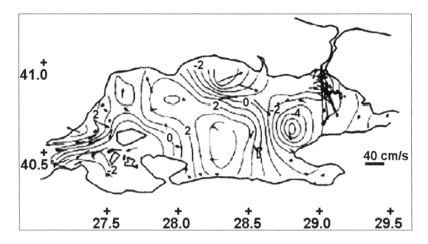

*Figure 3.* ADCP measured currents at 10 m superimposed on Dynamic Height Anomaly during March 1992 (Beşiktepe *et al.*, 1994). The Bosphorus jet and Dardanelles with high velocities was clearly identified.

## 1.3. EASTERN MEDITERRANEAN SEA

Turkey has a long coastline along the northern periphery of the Eastern Mediterranean Sea, which is a temperate semi-enclosed basin lying to the east of the 26 qE longitude. It comprises two major sub-basins; the Aegean and Levantine Seas. Its most important thermohaline characteristics is the Levantine Intermediate Water (LIW) formed by convective mixing of surface waters up to depths of 300–500 m due to winter cooling and excessive evaporation. Surface water temperature reaches ~30°C and salinity of 39 ppt due to strong solar radiation and high evaporation. At the surface, the eastward flowing Atlantic water is continuously modified and looses its identity when it reaches to the northeastern Turkish coastal waters.

Multiple scales of interactions define the general circulation of the Eastern Mediterranean. A complex system of mesoscale eddies; jets and patches of water masses embedded in the general circulation are the predominant characteristic features (Tzipernan and Malanote-Rizzoli, 1991). The Levantine Basin circulation consists of sub-basin-scale gyres driven by wind and thermohaline forcing, jets feeding the gyres, and a number of embedded coherent eddies (Ozsoy *et al.*, 1989, 1993). The cyclonic Rhodes Gyre was recognized as one of the most dominant circulation feature between Crete and Cyrups (Robinson *et al.*, 1991).

Climatological wind field in the Eastern Mediterranean is dominated by westerlies in winter, and northwesterly Etesian winds in summer (Middelandse Zee, 1957; Brody and Nestor, 1980; May, 1982). The region is subject to frequent outbreaks of cold and dry air masses from the Anatolia in the north.

## 2. Tanker Traffic and Routes

During the 1990s, the Caspian Sea, Black Sea and Central Eurasia region emerged as one of the most important new sources of world oil supply, leading to high oil traffic in the area. Oil traffic is expected to increase due to these emerging sources of petroleum as well as increased local production in the region. Pipelines currently online from Russia, Georgia, Kazakhstan and Azerbaijan, supply crude oil to terminals at the Russian, Ukrainian and Georgian Black Sea coasts.

International Tanker Owners Pollution Federation Limited (IPIECA) work, "Oil Spill Preparedness Regional Initiative", indicates an increasing trend of oil transport in the region. Kazakstan's oil production is expected to be more than double by 2010. Russia's oil production is estimated to grow almost 50%. Azarbaijan's production is predicted to increase by five fold. Overall, crude oil and products handled and shipped from different facilities is expected to double by 2010 (IPIECA, 2005). About 155 million tonnes of crude oil is to be transported through the Turkish Straits in 2010, compared to 119 million tones in 2002. Ceyhan port, located at the Turkish Mediterranean coast, is another route for transferring Caspian oil through pipeline from Baku. It was completed on July 2006 and is currently operating at 1 million barrels per day, one tenth of its expected capacity. Therefore, an increase in its operational capacity would imply a considerable increase in tanker traffic for the Eastern Mediterranean Sea.

## 3. Oil Spill Accidents

Understanding the history of oil spills from tankers helps tackling the oil spill problem more appropriately. The oil spill data retrieved from

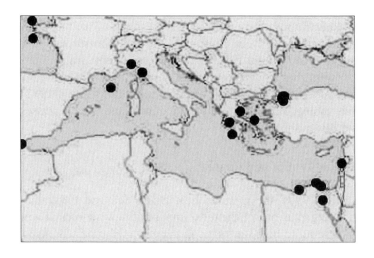

*Figure 4.* Major Tanker Spills (>700 t) since 1990. International Tanker Owners Pollution Federation Ltd., ITOPF (Martini and Patruno, 2005).

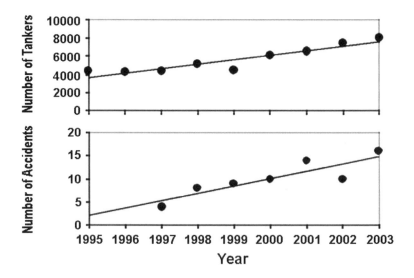

*Figure 5.* Trend Tanker Traffic in the Bosphorus (Ece, 2005) (top) and number of accidents in the Bosphorus (Korcak, 2005) (bottom).

International Tanker Owners Pollution Federation (ITOPF) and Regional Marine Pollution Emergency Response Centre for the Mediterranean Sea (REMPEC) databases indicate that approximately 80,000 t of oil have been spilled in the Mediterranean Sea between 1990 and 2005 (Martini and Patruno, 2005) at locations shown in Figure 4. Three major accidents that took place in the Bosphorus account for 45% of the total quantity spilled in Mediterranean Sea during this period.

The records suggest high risk of oil spills in the Bosphorus Strait, which is clearly supported by increasing tanker accidents in parallel with increasing tanker traffic (Figure 5). The most drastic tanker accidents were the INDEPENDENTA (1979), FAHIRE GUNERI (1984), JAMBUR (1990), NASSIA (1994), and VOLGONEFT-248 (1999). Vessel traffic in the 30 km-long Bosporus has grown to about 50,000 vessels per year with approximately 5,000 of these carrying oil or LNG.

## 4.    Status of Conventions and Contingency Plans

Turkey has ratified the major protocols related to oil pollution: 1990 International Convention on Oil Pollution Preparedness, Response and Cooperation (1990 OPRC); 1992 Protocol to the International Convention on Civil Liability for Oil Pollution Damage (1992 CLC) and the International Fund for Compensation for Oil Pollution Damage (1992 Fund). The status of Turkey in major protocols is shown in Table 1. By these protocols Turkey is obliged to undertake, individually or jointly, all appropriate measures in accordance with the provisions of these Protocols and Annexes to prepare for, and respond to, a pollution incident by hazardous and noxious substances. In addition to these international conventions and protocols, Barcelona (Mediterranean countries) and Bucharest (Black Sea countries) Conventions are the two regional conventions that the Republic of Turkey under-takes. Turkey ratified the Barcelona Convention (1976, amended in 1995) and its first Emergency protocol in 1981 and the new Emergency Protocol in 2003, which is called Protocol Concerning Cooperation in Preventing Pollution from Ships and, in Cases of Emergency, Combating Pollution of the Mediterranean Sea. The Convention on the Protection of the Black Sea Against Pollution, also referred to as "Bucharest Convention", was signed in Bucharest in April 1992. It is the basic framework of agreement and three specific Protocols, which

TABLE 1. The status of international protocols Ratification or Accession of conventions –Turkey (http://www.imo.org/).

| Convention | Deposit date | Date of entry into force | Status |
|---|---|---|---|
| CLC 69 | – | | |
| CLCPROT76 | – | | |
| CLC PROT 92 | 17/08/2002 | 17/08/2002 | ACCESSION |
| FUND 71 | – | | |
| FUND PROT 1976 | – | | |
| FUND PROT 1992 | 17/08/2001 | 17/08/2002 | ACCESSION |
| FUND PROT 2000 | – | | |
| HNS 96 | – | | |
| IMO AMEND-91 | – | | |
| IMO AMEND-93 | 04/05/2001 | 07/11/2002 | ACCEPTANCE |
| IMO CONVENTION 91 | – | | |
| IMO CONVENTION 91 | – | | |
| INTERVENTION 69 | – | | |
| INTERVENTION PROT | – | | |
| MARPOL ANNEX I/II | 10/10/1990 | 10/01/1991 | ACCESSION |
| MARPOL ANNEX III | – | | |
| MARPOL ANNEX IV | – | | |
| MARPOL ANNEX V | 10/10/1990 | 10/01/1991 | ACCEPTANCE |
| MARPOL ANNEX VI | – | | |
| OPRC 90 | 01/07/2004 | 01/10/2004 | ACCESSION |
| OPRC/HNS 2000 | – | | |
| SOLAS 74 | 31/07/1980 | 31/10/1980 | ACCESSION |
| SOLAS AGR96 | – | | |
| SOLAS PROT 78 | – | | |
| SOLAS PROT 88 | – | | |

are the control of land-based sources of pollution, dumping of waste, and joint action in the case of accidents (such as oil spills). The implementation of the Convention is managed by the Commission for the Protection of the Black Sea Against Pollution and its Permanent Secretariat in Istanbul, Turkey. Turkey ratified the convention in March 1994 (Table 2).

TABLE 2. The status of regional protocols Ratification or Accession of Conventions – Turkey.

| Convention | Deposit date | Date of entry into force | Status |
|---|---|---|---|
| BARCELONA | 16/02/1976 | 18/09/2002 | ACCESSION |
| BUCHAREST | 21/04/1992 | 14/04/1994 | ACCEPTANCE |

Implementation of these agreements and their associated resources and networks presents a great opportunity to move the region towards international standards of spill readiness and claims settlement.

Turkey is in the stage of drafting the National Oil Spill Contingency Plans (NCP). In order to support the legislation studies, a new project was prepared by Turkish Undersecretariat of Maritime Affairs to carry out an analysis towards constructing a national spill-response plan, which will improve the ability of the region to respond effectively in the event of an oil spill and contribute to the region's environmental protection.

There are some governmental and private organizations in Turkey with oil spill equipment, but with limited capacity to respond to oil loading/discharge facilities. They have the same first response (tier 1) capabilities and lack access and procedures for escalating to higher tier response. There is some government owned equipment that is operated by the coastal safety and ship salvage administration, and also by some contractors with oil spill response experience. A limited amount of oil spill clean up equipment is owned by oil companies in Turkey, located at main oil terminals. Terminals and oil companies have the best supplies of equipment.

# References

Armi, L. and Farmer, D.M., 1987, A generalization of the concept of maximal exchange in a strait. *Journal of Geophysical Research*, **83**, 873–883.

Besiktepe, S., 1991, Some Aspects of the Circulation and Dynamics of the Sea of Marmara. Ph.D. thesis, Middle East Technical University, Erdemli, Icel.

Besiktepe, S., Ozsoy, E. and Unluata, U., 1993, Filling of the Sea of Marmara by the Dardanelles Lower Layer Inflow. *Deep-Sea Research,* **40**, 1815–1838.

Beşiktepe, S., Sur, H.I., Ozsoy, E., Latif, M.A., Oguz, T. and Unluata, U., 1994, The Circulation and Hydrography of the Marmara Sea, *Progress in Oceanography*, 34, 285–334.

Brody, L.R. and Nestor, M.J.R., 1980, *Regional Forecasting Aids for the Mediterranean Basin,* Handbook for Forecasters in the Mediterranean, Part 2. Naval Environmental Prediction Research Facility, Monterey, California, Technical Report TR 80-10, 178 pp.

De Filippi, G.L., Iovenitti, L. and Akyarlı, A., 1986, Current analysis in the Marmara-Bosphorus junction. *1st AIOM (Assoclazione di Ingegneria Ofshore e Marina) Congress,* Venice, June 1986, pp. 5–25.

Ece, N.J., 2005, The Accident Analysis of the Strait of Istanbul from the Points of Safety Navigation and Environment and Evaluation of Innocent Passage. Ph.D. thesis, Gazi University, Ankara.

Farmer, D.M. and Armi, L., 1986, "Maximal two-layer exchange over a sill and through the combination of a sill and contraction with barotropic flow. *Journal of Fluid Mechanics,* **164**, 53–76.

Gawarkiewicz, G., Korotaev, G., Stanichny, S., Repetin, L. and Soloviev, D., 1999, Synoptic upwelling and cross-shelf transport processes along the Crimean coast of the Black Sea. *Continental Shelf Research,* **19**, 977–1005.

IPIECA, Briefing paper, 2005, Oil Spill Preparedness Regional Initiative (Caspian Sea – Black Sea – Central Eurasia) OSPRI (http://www.ipieca.org).

Korotaev G., Oguz, T., Nikiforov, A. and Koblinsky, C., 2003, Seasonal, interannual and mesoscale variability of the Black Sea upper layer circulation derived from altimeter data. *Journal of Geophysical Research,* **108**(C4), 3122.

Korcak M., 2005, Experiences of Oil Spill Accidents and New Legislation Studies in Turkey. *Satellite Monitoring and Assessment of Sea-based Oil Pollution in the Black Sea,* Istanbul, Turkey, 13–15 June 2005.

Krivosheya, V.G., Titov, V.B., Ovchinnikov, I.M., Kosyan, R.D., and Skirta, A.Y., 2000, The influence of circulation and eddies on the depth of the upper boundary of the hydrogen sulfide zone and ventilation of aerobic waters in the Black Sea. *Oceanology* (Eng. Transl.), **40**,767–776.

Krivosheya, V.G., Ovchinnikov, I.M., and Skirta, A. Yu., 2002, *Interannual variability of the cold intermediate layer renewal in the Black Sea.* In: Multidisciplinary investigations of the northeast part of the Black Sea, A.G. Zatsepin and M.V. Flint, eds, Nauka, Moscow, pp. 27–39.

LMIU 2001. *LMIUShipping Data,* Lloyd's Marine Intelligence Unit, 2001.

Middelandse Zee 1957. "Oceanographic and meteorological data", *Nederlands Meteorologisch Institut,* 91 pp.

May, P.W., 1982, *Climato Logical Flux Estimates of the Mediterranean Sea,* Part I: Winds and Wind Stresses. Report 54, NORDA, NSTL Station, 56 pp. Publication, 238 pp.

Martini, N. and Patruno, A.R., 2005, Oil Pollution Risk Assessment and Preparedness in the East Mediterranean. *International Oil Spill conference.* Miami, USA, 15–19 May 2005.

Oguz, T. and Sur, H.I., 1989, A two-layer model of water exchange through the Dardanelles Strait. *Oceanology Acta,* **12,** 23–31.

Oguz, T., Ozsoy, E., Latif, M.A., Sur, H.I. and Unluata, U., 1990a, Modelling of hydrauliing of hydraulically controlled exchange flow in the Bosphorus Strait. *Journal of Physical Oceanography,* **20**(7), p.945–965.

Oguz, T., Latif, M.A., Sur, H.I., Ozsoy, E. and Unluata, U., 1990b, *On the dynamics of the southern Black Sea.* In: The Black Sea Oceanography, J. Murry and E. Izdar, eds, NATO/ASI Series, Kluwer, Dordrecht, pp. 43–64.

Oguz, T., Aubrey, D.G., Latun, V.S., Demirov, E., Koveshnikov, L., Sur, H.I., Diacanu, V., Besiktepe, S., Duman, M., Limeburner, R. and Eremeev, V.V., 1994, Mesoscale circulation and thermohaline structure of the Black sea observed during HydroBlack'91. *Deep-Sea Research,* **I41,** 603–628.

Oguz, T., Ivanov, L.I. and Besiktepe, S., 1998, *Circulation and hydrographic characteristics of the Black Sea during July 1992.* In: L. Ivanov and T. Oguz, eds, Ecosystem Modeling as a Management Tool for the Black Sea, NATO ASI Series, Environmental Security – vols. 47, 2. Kluwer, Dordrecht, pp. 69–92.

Oguz, T. and Besiktepe, S., 1999, Observations on the Rim Current structure, CIW formation and transport in the western Black Sea. *Deep- Sea Research,* **I46,** 1733–1753.

Oguz, T., 2005a, Hydraulic adjustment of the Bosphorus exchange flow". *Geophyical Research Letters,* **32,** L06604, doi: 10.1 029/2005GL022353.

Oguz, T., Tugrul, S., Kideys, A.E., Ediger, V., Kubilay, N., 2005b, *Physical and biogeochemical characteristics of the Black Sea.* The Sea, Vol. **14,** Chapter 33, pp. 1331–1369.

Ozsoy E., Oguz, T., Latif, M.A. and Unluata, U., 1986, *Oceanography of the Turkish Straits.* First Annual Report, Institute of Marine Sciences, Middle East Technical University, Vol. 1, 269 pp.

Ozsoy, E., Unluata, U. and Top, Z., 1993, The Mediterranean water evolution, material transport by double diffusive intrusions, and interior mixing in the Black Sea. *Progress in Oceanography,* **31,** 275–320.

Ozsoy, E. and Unluata, U., 1997, Oceanography of the Black Sea: a review of some recent results. *Earth Science Review,* **42,** 231–272.

Ozsoy, E., Latif, M.A. and Beşiktepe, S., 2002, The Current System of the Bosphorus Strait Based on Recent Measurements, *The 2nd Meeting on the Physical Oceanography of Sea Straits,* Villefranche, 15–19 April 2002, pp. 177–180.

Robinson, A.R., Golnaraghi, M., Leslie, W.G., Artegiani, A., Hecht, A., Lazzoni, E., Michelato, A., Sansone, E., Theocharis, A. and Unluata, U., 1991, The Structure and Variability of the Eastern Mediterranean general circulation. *Dynamics of the Atmosphere and Oceans*, **15**, 215–240.

Tzipernan, E. and Malanote-Rizzoli, P., 1991, The Climatological Seasonal Circulation of the Mediterranean Sea. *Journal of Marine Research,* **49**,411–434.

Unluata, U., Oguz, T., Latif, M.A. and Ozsoy, E., 1990, *On the physical oceanography of the Turksh Straits.* In The Physical Oceanography of the Sea Straits, L.J. Pratt, Ed., NATO/ASI Series, V. 318, pp. 25–60, Kluwer, Dordrecht.

# UK: RECENT COUNTER POLLUTION R&D ACTIVITIES

K. COLCOMB[†]

*Maritime and Coastguard Agency\Senior Scientist*
*Counter Pollution & Response, 105 Commercial Road,*
*Southampton, SO15 1EG, United Kingdom*

**Abstract**[*]. The UK Maritime and Coastguard Agency commission research in all areas of its business, primarily: "safer lives, safer ships and cleaner seas". The current annual R&D budget stands at approximately $1.5 m. Much MCA led research carried out for counter pollution activity is multi-agency collaborative work attracting funds from both the UK public and private sector. The primary objective of this work is to improve the UK's position to plan for and respond to oil and chemical maritime pollution. Key recent projects include the determination of the limiting viscosity for the use of dispersants in oil spill response; the outcome of this work underpins the MCA dispersant response philosophy in the light of increasing levels of heavy oils transiting UK waters. A comprehensive investigation into the potential ecological effects of the use of dispersants was concluded in 2004. The UK coastal and marine resource atlas, a GIS platform capturing 100 layers of environmental sensitivities, provides planners with a comprehensive reference tool. New projects recently approved by the MCA Research Committee include: A new GIS platform to present potentially polluting wrecks in UK waters, a very heavy fuel oil risk assessment, a risk assessment for Hazardous and Noxious Substances in UK waters, the design of a waste treatment infrastructure for dealing with significant quantities of oily waste, and the development of a standard shoreline clean-up assessment reporting protocol.

**Keywords:** oil spill response, counter pollution, dispersant

---

[†] To whom correspondence should be addressed. E-mail: kevin_colcomb@mcga.gov.uk
[*] Full presentation available in PDF format on CD insert.

W. F. Davidson, K. Lee and A. Cogswell (eds.), *Oil Spill Response: A Global Perspective.* 341
© Springer Science + Business Media B.V. 2008

# ICE OVERVIEW FOR THE GULF OF ST. LAWRENCE AND THE ST. LAWRENCE RIVER

M. BLOUIN[†] & E. VAILLANT
*Canadian Coast Guard, Fisheries and Oceans Canada,*
*Canadian Ice Service, Environment Canada, 101*
*Champlain Blvd., Quebec City, G1K 7Y7, Canada*

**Abstract.** Through the years the Canadian Coast Guard and Environment Canada have developed an expertise of working in the ice environment and in characterizing the types of ice. Assisting the marine industry in ice is quite a challenge because of the dynamics of ice, the air temperature and the water temperature. During this presentation I gave a brief overview of the types of ice and the dynamics of ice and I discussed the icebreaking services provided in the St. Lawrence River and in the Gulf.

**Keywords:** ice dynamics, icebreaking

---

[†] To whom correspondence should be addressed. E-mail: blouinm@dfo-mpo.gc.ca

W. F. Davidson, K. Lee and A. Cogswell (eds.), *Oil Spill Response: A Global Perspective.*    343
© Springer Science + Business Media B.V. 2008

# LIST OF PARTICIPANTS (NATO ARW)

## "Third Annual Workshop on Oil Spill Response"

## Dartmouth, Nova Scotia, Canada, 11–13 October 2006

### Key Speakers from NATO Countries:

Martin Blouin

Fisheries and Oceans Canada,
Coast Guard, 101 Champlain Blvd,
Québec, G1K 7Y7, Canada
Tel: (418) 648-4557
Fax: (418) 648-4003
E-mail: blouinm@dfo-mpo.gc.ca

Michel C. Boufadel

Temple University, Department of Civil
and Environmental Engineering,
Room 511, Temple University,
1947 N. 12th Street, PA 19122, USA
Tel: (215) 204-7871
Fax: (215) 204-4696
E-mail: boufadel@temple.edu

Per Johan Brandvik

SINTEF, Material and Chemistry,
Marine Environmental,
Trondheim, N-7465, Norway
Tel: +47 909 58 576
E-mail: Per.Brandvik@sintef.no

Ian Buist

SL Ross Environmental Research,
Suite 200, Ottawa, ON K1G 0Z4,
Canada
Tel: (613) 232-1564
Fax: (613) 232-6660
E-mail: ian@slross.com

Zhi Chen

Concordia University
1455 de Maisonneuve Blvd. W.
Montreal, QC H3G 1M8, Canada
Tel: (514) 848 2424 x 8775
E-mail: zhichen@bcee.concordia.ca

Tülay Çokacar

Tubitak MRC, Chemistry
and Environment Institute,
P.O. Box. 21 41470, Gebze,
Kocaeli, Turkey
Tel: +90 262 677 29 37
E-mail: Tulay.Cokacar@mam.gov.tr

Kevin Colcomb

Maritime and Coastguard Agency\Senior
Scientist Counter Pollution & Response,
105 Commercial Road, Southampton,
SO15 1EG, United Kingdom
Tel: +44(0)23 8032 9100
Fax: +44 (0)23 8032 9485
E-mail: kevin_colcomb@mcga.gov.uk

Simon Courtenay

Fisheries and Oceans Canada at the,
Canadian Rivers Institute,
Department of Biology,
University of New Brunswick,
Bag Service #45111, 10 Bailey Drive,
Fredericton, NB E3B 6E1, Canada
Tel: (506) 451-6892
Fax: (506) 453-3583
E-mail: courtenays@dfo-mpo.gc.ca

Walter Davidson

NATO/CCMS National Representative
Director (National Facilities) National
Research Council, Room 1025,
Ottawa, ON K1A 0R6, Canada
Tel: (613) 990 0914
Fax: (613) 993 4291
E-mail: walter.davidson@nrc-cnrc.gc.ca

Bernard Doyon

Fisheries and Oceans Canada,
Hydraulic Engineering Division,
Room: 146
101 Champlain Boulevard,
Mail Stop: QBC, Québec,
G1K 7Y7, Canada
Tel: (418) 648-3783
Fax: (418) 649-6201
E-mail: DoyonB@dfo-mpo.gc.ca

John French

Prince William Sound Regional Citizens
Advisory Council\Director,
P.O. Box 1470, 99664, Seward,
Alaska, USA
Tel: (907) 224-4429
E-mail: jsfrench@arctic.net

Deborah French-McCay

70 Dean Knauss Drive, Narragansett,
02882-1143, RI, USA
Tel: (401) 789-6224
Fax: (401) 798-1932
E-mail: dfrench@appsci.com

Ron Goodman

Innovative Ventures Ltd., P.O. Box 670
Cochrane, AB T4C 1A8, Canada
Tel: (403) 932-4331
Fax: (403) 932-4263
E-mail: goodmanr@cia.com

Charles Hannah

Fisheries and Oceans Canada,
Oceans Sciences Division,
Bedford Institute of Oceanography,
P.O. Box 1006, Dartmouth,
Nova Scotia, B2Y 4A2, Canada
Tel: (902) 426-5961
E-mail: HannahC@dfo-mpo.gc.ca

Peter Hodson

Queen's University,
116 Barrie Street, Kingston,
ON K7L 3N6, Canada
Tel: (613) 533 -6129
E-mail: hodsonp@biology.queensu.ca

Bruce Hollebone

Environment Canada, Oil Research
Section, Ottawa, ON K1A 0H3, Canada
Tel: (613) 991-4568
Fax: (613) 991-9485
E-mail: bruce.hollebone@ec.gc.ca

Thierry Jacques

Unité de Gestion du Modèle
Mathématique Mer du Nord,
(UGMM/MUMM) Institut Royal des
Sciences Naturelles de Belgique,
Gulledelle, 100 B-1200 Bruxelles,
Belgium
Tel: +32 (0)2 773 21 24
Fax: +32 (0)2 770 69 72
E-mail: t.jacques@mumm.ac.be

Hanne Greiff Johnsen

Senior research scientist
Environmental Technology
Statoil Research Centre
Arkitekt Ebbellsvei 10, Rotvoll
N-7005 Trondheim, Norway
Tel: +47 73584011
Fax: +4773967286
E-mail: hanjo@statoil.com

Ken Lee

Centre of Offshore Oil and Gas
Environmental Research,
Bedford Institute of Oceanography,
P.O. Box 1006, Dartmouth,
NS B2Y 4A2, Canada
Tel: (902) 426-7344
Fax: (902) 426-1440
E-mail: LeeK@mar.dfo-mpo.gc.ca

Zhengkai Li

Centre for Offshore Oil and Gas,
Environmental Research,
Bedford Institute of Oceanography,
1 Challenger Drive, P.O. Box 1006,
Dartmouth, NS B2Y 4A2, Canada
Tel: (902) 426-3442
Fax: (902) 426-1440
E-mail: LiZ@mar.dfo-mpo.gc.ca

| | |
|---|---|
| Qianxin Lin | Louisiana State University/Wetland, Biogeochemistry Institute, Baton Rouge, 70803, LA, USA<br>Tel: (225) 578-8889<br>Fax: (225) 578-6423<br>E-mail: comlin@lsu.edu |
| Johan Marius Ly | Norwegian Coastal Administration, Department of Emergency Response, P.O. Box 125, NO-3191 Horten, Norway<br>Tel: +47 33 03 48 00<br>Fax: +47 33 03 49 49<br>E-mail: johan-marius.ly@kystverket.no |
| Jim Mackey | LAMOR Corporation LLC, 28045 Ranney Parkway, Unit G, Cleveland, OH 44145, USA<br>Tel: (440) 871-8000 ext.156<br>Fax: (440) 871-8104<br>E-mail: jim.mackey@lamor.com |
| Dr. Etienne Mansard | Canadian Hydraulics Centre/National Research Council, 1200 Montreal Rd., M-32, Ottawa, ON K1A 0R6, Canada<br>Tel: (613) 993-2417<br>Fax: (613) 952-7679<br>E-mail: etienne.mansard@nrc-cnrc.gc.ca |
| François Merlin | CEDRE, 715 r. A. Colas, CS:41836 Brest Cedex 2, F-29218, France<br>Tel: +33 2 9833 6706<br>Fax: +33 2 98 44 91 38<br>E-mail: francois.merlin@cedre.fr |
| Joseph Mullin | MMS/Engineering and Research Branch MS-4021, Herndon, VA 20170-4817, USA<br>Tel: (703) 787-1556<br>Fax: (703) 787-1549<br>E-mail: Joseph.Mullin@mms.gov |

Tim Nedwed                ExxonMobil, Offshore Division, 3319
Mercer (Delivery), Houston,
TX 77027-2189, USA
Tel: (713) 431-6923
Fax: (713) 431-6423
E-mail: Tim.j.nedwed@exxonmobil.com

James Payne                Payne Environmental Consultants, Inc.
1991 Village Park Way, Suite 206 B,
Encinitas, CA 92024, USA
Tel: (760) 942-1015
Fax: (760) 942-1036
E-mail: jrpayne@sbcglobal.net

Roger Percy                Environment Canada, 16th Floor, Queen
Square, Dartmouth, NS B2Y 2N6, Canada
Tel: (902) 426-2576
Fax: (902) 426-9709
E-mail: roger.percy@ec.gc.ca

Marko Perkovic           University of Ljubljana, Faculty of
Maritime Studies and Transportation,
Pot pomorscakov 4, SI-6320 Portoroz,
Slovenia
Tel: +386 5 67-480-90
Fax: +386 5 67-480-91
E-mail: Marko.Perkovic@fpp.uni-lj.si

Steve Potter                SL Ross Environmental Research,
Suite 200, 717 Belfast Rd., Ottawa,
ON K1G 0Z4, Canada
Tel: (613) 232-1564
E-mail: steve@slross.com

Ivar Singsaas              SINTEF Materials and Chemistry,
Brattorkaia 17, N-7465 Trondheim
Norway
Tel: +47 +47 982 43 467
Fax: +47 + 47 930 70 026
E-mail: ivar.singsaas@sintef.no

Lauri Solsberg

Counterspil Research Inc.
205 – 1075 West 1st Street
North Vancouver, BC V7P 3T4
Canada
Tel: (604) 990-6944
Fax: (604) 990-6945
E-mail: mail@counterspil.com

Stein Erik Sorstrom

Senior Adviser, International Operations
SINTEF Technology and Society,
S P Andersens v 5, 7465 Trondheim,
Norway
Tel: +47 99536050
E-mail: Stein.E.Sorstrom@sintef.no

Darren Trites

DSS Marine
71 Wright Ave
Dartmouth, NS B3B 1H4
Canada
Tel: (902) 835-4848
Fax: (902) 835-6269
E-mail: dtrites@dssmarine.com

Andrew Tucci

United States Coast Guard\Office of
Response Field Activities Directorate
United States Coast Guard\Commandant
(G-MOR), 2100 2nd St. S.W,
Washington, DC 20593-0001, USA
Tel: (202) 372-2234
E-mail: Andrew.E.Tucci@uscg.mil

Al Venosa

U.S. Environmental Protection Agency
26 W. Martin Luther King Drive,
Cincinnati, OH 45268, USA
Tel: (513) 569-7668
Fax: (513) 569-7105
E-mail: venosa.albert@epa.gov

| Zhendi Wang | Environment Canada, Emergencies Science & Technology Division, 335 River Road Ottawa, ON K1A 0H3, Canada Tel: (613) 990-1597 Fax: (613) 991-9485 E-mail: Zhendi.Wang@ec.gc.ca |

## Other Participants from NATO Countries

| Yuri Bobrov | NATO Interpreter, 10. av. PDT Robert Schuman, Strasbourg, 67000, France Tel: +33 6 80 66 25 96 Fax: +33 3 88 35 41 29 E-mail : bobrov@noos.fr |
| Jay Bugden | Centre for Offshore Oil and Gas, Environmental Research, 1 Challenger Drive, P.O. Box 1006, Dartmouth, NS B2Y 4A2, Canada Tel: (902) 426-7256 E-mail : BugdenJ@mar.dfo-mpo.gc.ca |
| Les Burridge | Marine Environmental Sciences Division Main Building – Floor: 2nd – Room: 2-37 St Andrews Biological Station, 531 Brandy Cove Road, St Andrews, NB E5B 2L9 Canada Tel: (506) 529-5903 E-mail: BurridgeL@mar.dfo-mpo.gc.ca |
| Steve "Vinnie" Catalano | Cook Inlet Regional Citizens' Advisory Council, 910 Highland Avenue, Kennai, 99611, AK, USA Tel: (907) 283-7222 Fax: (907) 283-6102 E-mail: williams@circac.org |

Hermanis Cernovs

Republic of Lativa, Latvian Naval
Forces
Meldru str 5A, Riga, LV-1015, Latvia
Tel: 3717082061
Fax: 3717353215
E-mail: cernovs@mrcc.lv

Susan Cobanli

Centre for Offshore Oil and Gas,
Environmental Research, Bedford
Institute of Oceanography, 1 Challenger
Drive, P.O. Box 1006, Dartmouth, NS
B2Y 4A2 Canada
Tel: (902) 426-2479
Fax: (902)-426-1440
E-mail: CobanliS@mar.dfo-mpo.gc.ca

Andrew Cogswell

Centre for Offshore Oil and Gas,
Environmental Research (COOGER),
Bedford Institute of Oceanography, 1
Challenger Drive, P.O. Box 1006,
Dartmouth, NS B2Y 4A2, Canada
Tel: (902) 426-1438
Fax: (902) 426-1440
E-mail:Cogswella@mar.dfo-mpo.gc.ca

Melinda Cole

Centre for Offshore Oil and Gas,
Environmental Research, 1 Challenger
Drive, PO Box 1006, Dartmouth, NS
B2Y 4A2, Canada
Tel: (902) 426-4172
E-mail: colem@mar.dfo-mpo.gc.ca

Mike Comeau

Maritime Forces Atlantic (MARLANT)
Formation Safety and
Environment/Maritimes Forces, P.O.
Box 99000, Station Forces, Halifax, NS
B3K 5X5, Canada
Tel: (902) 721-1122
Fax: (902) 721-5417
E-mail: Comeau.MJ3@forces.gc.ca

| | |
|---|---|
| Darin Connors | Eastern Canada Response Corporation Ltd.<br>481 Polymoore Drive, Corunna, ON N0N 1G0, Canada<br>Tel: (519) 862-2281<br>Fax: (519) 862-3510<br>E-mail: dconnors@ecrc.ca |
| David Cooper | SAIC Canada, 335 River Road, Ottawa, ON K1A 0H3, Canada<br>Tel: (613) 991-1841<br>Fax: (613) 991-1673<br>E-mail: david.cooper@saiccanada.com |
| Andre d'Entremont | Chevron Canada Resources, 500 – Fifth Avenue S.W., Calagary, AB T2P OL7 Canada<br>Tel: (403) 234-5294<br>Fax: (403) 234-5947<br>E-mail : adentremont@chevrontexaco.com |
| Mark Dalton | Environment Canada, Environmental Protection Branch, 45 Alderney Dr. Queen's Sq., Dartmouth, NS BY 2N6, Canada<br>Tel: (902) 426-7052<br>Fax.: (902) 426-7924<br>E-mail: daltonmarkec@dfo-mpo.gc.ca |
| Leigh Dehaven | US Environmental Protection Agency 1200 Pennsylvania Avenue NW, Mail Code 5104A, 20460, Washington, DC USA<br>Tel: (202) 564-1974<br>Fax: (202) 564-2625<br>E-mail: dehaven.leigh@epa.gov |

Ian Denness

ConocoPhillip Canada\Loss Prevention
Coordinator, 401 – 9th Avenue S.W
Calgary, AB T2P 2H7, Canada
Tel: (403) 233-3894
E-mail:
Ian.Denness@concocophillips.com

Jennifer Dixon-Mason

Centre for Offshore Oil and Gas
Environmental Research, Bedford
Institute of Oceanography, 1 Challenger
Drive, P.O. Box 1006, Dartmouth, NS
B2Y 4A2, Canada
Tel: (902) 426-5467
Fax: (902) 426-1853
E-mail:DixonJA@mar.dfo-mpo.gc.ca

Dr. Bekmamat Djenbaev

Biology-Soil Institute/NAS of Kyrgyz
Rep, Ave. Chui, Bishkek, 720071,
Kyrgyz Republic
Tel: +996-312-243 991
Fax: +996-312-243 607
E-mail: bekmamat2002@mail.ru

Marie-France Gauthier

Canadian Ice Service, Environment
Canada, 373 Sussex Drive, E-3, Ottawa,
ON K1A 0H3, Canada
Tel: (613) 943-8026
Fax: (613) 996-4218
E-mail : Marie-rance.Gauthier@ec.gc.ca

Terri Green

Environment Canada/Senior
Communications Advisor, 16th Floor,
Queen Square, 45 Alderney Drive,
Dartmouth, NS B2T 1G4, Canada
Tel: (902) 426-9168
Fax: (902) 426-5340
E-mail: Terri.Green@ec.gc.ca

| | |
|---|---|
| Stephane Grenon | Environment Canada/Environmental, Emergencies Centre Saint-Laurent, 105 McGill Street, Montreal, QC H2Y 2E7, Canada<br>Tel: (514) 283-2345<br>E-mail : GrenonStephaneec@dfo-mpo.gc.ca |
| Ioulia Grigorieva-Maes | NATO, 19, Rue Roberts-Jones, B-1180, Brussels, Belgium<br>Tel: +32 2 347 19 00<br>E-mail: g.ioulia@belgacom.net |
| Trond Hansen | NorLense AS, Fishkebøl, N-8317 Strøstad, Norway<br>Tel: +47 76 11 81 80<br>Fax: +47 908 21 808<br>E-mail: trond@norlense.no |
| Kats Haya | Marine Environmental Sciences Division<br>St Andrews Biological Station, 531 Brandy Cove Road, St. Andrew's, NB E5B 2L9, Canada<br>Tel: (506) 529-5916<br>E-mail: HayaK@mar.dfo-mpo.gc.ca |
| Peter Hennigar | Environmental Emergencies/Environment Canada/Atlantic Region, 45 Alderney Drive, Dartmouth, NS B2Y 2N6, Canada<br>Tel: (902) 426-6191<br>Fax: (902) 426-9709<br>E-mail: peter.hennigar@ec.gc.ca |
| Peter Holgersen | Forsvarskommandoen, Defence Command Denmark, P.O. Box 202, 2950 Vedbaek, Denmark |

Tel: +45 4567 3141
Fax: +45 4567 3159
E-mail: seapol@mil.dk

Jamie Joudrey

Centre for Offshore Oil and Gas
Environmental Research, 1 Challenger
Drive, P.O. Box 1006, Dartmouth, NS
B2Y 4A2, Canada
E-mail: JoudreyJ@mar.dfo-mpo.gc.ca

Robert Keenan

Environment Canada/Environmental
Emergencies, 45 Alderney Drive,
Dartmouth, NS B2Y 2N6, Canada
Tel: (902) 426-1976
Fax: (902) 426-9709
E-mail: keenanrobertec@dfo-mpo.gc.ca

Gary Kennell

Environment Canada/Environmental
Protection Branch, 6 Bruce Street,
Mount Pearl, NF & Labrador, A1N 4T3,
Canada
Tel: (709) 772-2173
Fax: (709) 772-5097
E-mail: gary.kennell@ec.gc.ca

Paul Kepkay

Centre for Offshore Oil and Gas
Environmental Research, Bedford
Institute of Oceanography, 1 Challenger
Drive, P.O. Box 1006, Dartmouth, NS
B2Y 4A2, Canada
Tel: (902) 426-7256
E-mail : KepkayP@mar.dfo-mpo.gc.ca

Tom King

Centre for Offshore Oil and Gas
Environmental Research, Bedford
Institute of Oceanography, 1 Challenger
Drive, P.O. Box 1006, Dartmouth, NS
B2Y 4A2, Canada
Tel: (902) 426-4172
Fax: (902) 426-1853
E-mail: KingT@mar.dfo-mpo.gc.ca

Andrew Kinley

NATO, Rue Omer Lepreux 38, B-1081
Brussels, Belgium
Tel: +32 2 410 4229
Fax: +32 2 410 4229
E-mail: kinhon@scarlet.be

Nancy Kinner

University of New Hampshire, 35
Colovos Road, Durham, 03824, NH,
USA
Tel: (603) 862-1422
Fax: (603) 862-3957
E-mail: nancy.kinner@unh.edu

R.J. Kopchak

Oil Spill Recovery Institute, Cordova
Alaska, USA
E-mail: ecotrust@ak.net

Yves Lanthier

Environment Canada, 351 St Joseph
Boulevard Gatineau, Quebec, K1A 0H3,
Canada
Tel: (819) 997–8069
E-mail : yves.lanthier@ec.gc.ca

Joe LeClair

Fisheries and Oceans Canada, P.O. Box
1000, Dartmouth CCG Base, 2nd Floor,
Dartmouth, NS B2Y 4T3, Canada
Tel: (902) 426-3699
Fax: (902) 426-4828
E-mail: leclairj@mar.dfo-mpo.gc.ca

Nathalie Lowry

Environmental Emergencies Program –
Yukon/Environment Canada –
Environmental Protection Operations,
1782 Alaska Hwy, Whitehorse, YT Y1A
5B7, Canada
Tel: (867) 667-3405
Fax: (867) 667-7962
E-mail : Nathalie.Lowry@ec.gc.ca

| | |
|---|---|
| Xiaowei Ma | Centre for Offshore Oil and Gas Environmental Research, 1 Challenger Drive, P.O. Box 1006, Dartmouth, NS B2Y 4A2, Canada<br>Tel: (902) 426-3442<br>Fax: (902) 426–1440<br>E-mail : MaX@mar.dfo-mpo.gc.ca |
| Vincent Martin | Eastern Canada Response Corporation Ltd.<br>110 Montee-Calixa-Lavalle, Vercheres, QC J0L 2R0, Canada<br>Tel: (450) 583-5588<br>Fax: (450) 583-5414<br>E-mail: vmartin@simec.ca |
| John McKim | POL-E-MAR Inc., Ottawa, ON K2E 7K3 Canada<br>Tel:(613) 723-1541\(902) 466-2151<br>Fax: (613) 723-8692<br>E-mail: johnm@polemar.com |
| Reid McLeod | Eastern Canada Response Corporation Ltd.<br>Woodside Industrial Park, Dartmouth, NS, Canada<br>Tel: (902) 461-9170<br>Fax: (902) 461-9590<br>E-mail: RMcLeod@ecrc.ca |
| Amy Merten | NOAA, 7600 Sandpoint Way, NE, Seattle, WA 98115, USA<br>Tel: (206) 526-6829<br>E-mail: Amy.Merten@noaa.gov |
| Scott Miles | Louisiana State University, 1002I Energy, Coast & Environment Building, Baton Rouge, LA 70803, USA<br>Tel: (225) 578-4295<br>E-mail: msmiles@lsu.edu |

Maureen Murphy

One Ocean, Fisheries and Marine
Institute of Memorial University,
P.O. Box 4920
St. John's, NL A1R 5R3, Canada
Tel: (709) 778-0511
E-mail: maureen.murphy@mi.mun.ca

William "Nick" Nichols

U.S. EPA, Office of Emergency
Management/Regulatory and Policy
Development Division, Ariel Rios North
1200 Pennsylvania Ave., Washington,
DC 20460 (5104A), USA
Tel: (202) 564-1970
Fax: (202) 564-2625
E-mail: nichols.nick@epa.gov

Sue Olsen

Emergency Preparedness and Response
Public Health Agency of Canada,
Alberta/Northwest Territories Region,
9700 Jasper Avenue, Edmonton, AB
T5J 4C3, Canada
Tel: (780) 495-8785
Fax: (780) 495-5537
E-mail: olsensuehc@dfo-mpo.gc.ca

Sigurd Pacher

Deputy Head of Mission, Austrain,
Embassy, 445 Wilbrod Street, Ottawa,
ON K1N 6M7, Canada
Tel: (613) 789-1444
Fax: (613) 789-3431
E-mail: sigurd.pacher@bmaa.gv.at

Roger Percy

Environment Canada, 16th Floor, Queen
Square, Dartmouth, NS B2Y 2N6,
Canada
Tel: (902) 426-2576
Fax: (902) 426-9709
E-mail: roger.percy@ec.gc.ca

Andry W. Ratsimandresy     NSERC visiting scientist, Biol. and
                           Phys. Oceanography Section, Fisheries
                           and Oceans Canada NAFC, East White
                           Hills Rd., P.O. Box 5667 St. John's, NL,
                           A1C 5X1, Canada
                           Tel: (709) 772-5880
                           Fax: (709) 772-8138
                           E-mail: andry@ratsimandresy.org

Marek Rezko                Maritime Search & Rescue Service,
                           81-340 Gdynia, P.O. Box 375, Poland
                           Tel: +48 58 660 76 13
                           Fax: +48 58 660 76 14
                           E-mail: marek.reszko@sar.gov.pl

Alfonso Ruiz de Lobera     Jefe de Area de Relaciones
                           Internacionales y Estudios/Salvamento
                           Maritimo, Ministerio de Fomento, Spain
                           Tel: +34 917559126
                           Fax: +34 917559109
                           E-mail: interoper@sasemar.es

Debra Simecek-Beatty       NOAA, 7600 Sand Point Way NE,
                           Seattle, WA 98115, USA
                           E-mail:
                           Debra.Simecek-Beatty@noaa.gov

Mike Sinclair              Bedford Institute of Oceanography, 1
                           Challenger Drive, P.O. Box 1006,
                           Dartmouth, NS B2Y 4A2, Canada
                           Tel: (902) 426-3490
                           Fax: (902) 426-8484
                           E-mail : sinclairm@mar.dfo-mpo.gc.ca

Norm Snow                  Joint Secretariat/Executive Director
                           P.O. Box 2120, Inuvik, NT X0E 0T0,
                           Canada
                           Tel: (867) 777-2828
                           Fax: (867) 777-2610
                           E-mail: execdir@jointsec.nt.ca

Steve Solomon          Natural Resources Canada, Geological Survey of Canada (Atlantic), 1 Challenger Drive, P.O. Box 1006, Dartmouth, NS B2Y 4A2, Canada
Tel: (902) 426-8911
Fax: (902) 426-4104
E-mail: ssoloman@NRCan.gc.ca

Peter Thamer          Centre for Offshore Oil and Gas, Environmental Research, 1 Challenger Drive, P.O. Box 1006, Dartmouth, NS B2Y 4A2, Canada
Tel: (902) 426–4172
E-mail:ThamerP@mar.dfo-mpo.gc.ca

David Tilden          Hazardous Materials Specialist & Chairman, Artic Regional, Environmental Emergencies Team\Environment Canada, Prairie & Northern Region, Environment Protection Operations Division, Addr#301 5204-50th Avenue, Yellowknife, NT X1A 1E2, Canada
Tel: (867) 669-4728
Fax: (867) 873-8185
E-mail: david.tilden@ec.gc.ca

Christian Varescon          Total Fluides, Développement & Marketing, 51 Av. du Général de Gaulle LaDéfense 10, Paris La Défense Cedex F-92069 Paris La Défense, France
Tel: + 33 (0) 1 41 35 59 83
Fax: + 33 (0) 1 41 35 51 34
E-mail: christian.varescon@total.com

Peter Velez          Shell International Exploration & Production Company, 200 North Dariy Ashford, Houston, TX 77079, USA

|                    |                                                                                                                                                    |
|--------------------|----------------------------------------------------------------------------------------------------------------------------------------------------|
|                    | Tel: (281) 544-3000<br>Fax: (281) 544-2010<br>E-mail: peter.velez@shell.com                                                                        |
| Sue Williamson     | NATO/Threats & Challenges Section<br>Public Diplomacy Division/Assistant<br>Finance, Brussels, Belgium<br>Tel: +32-2 707 4928<br>E-mail: science.admin@hq.nato.int |
| Kari Workman       | Centre for Offshore Oil & Gas,<br>Environmental Research, 1 Challenger<br>Drive, P.O. Box 1006, Dartmouth, NS<br>B2Y 4A2, Canada<br>Tel: (902) 426-1436<br>Fax: (902) 426-1440<br>E-mail : WorkmanK@mar.dfo-mpo.gc.ca |
| Gina V. Ytteborg   | Shell Technology Norway AS/Cleaner<br>Production Team, Drammensv. 147A,<br>N-0277 Oslo, P.O. Box 1154, Sentrum,<br>Oslo, N-0107, Norway<br>Tel: +47 22 66 54 92<br>Fax: +47 22 66 53 41<br>E-mail: Gina.Ytteborg@shell.co |
| William Yeung      | Centre for Offshore Oil and Gas<br>Environmental Research, 1 Challenger<br>Drive, PO Box 1006, Dartmouth, NS<br>B2A 4A2, Canada<br>Tel: (902) 426-3258<br>E-mail: YeungW@mar.dfo-mpo.gc.ca |

## Key Speakers from Partner Countries

|              |                                                                                                                         |
|--------------|-------------------------------------------------------------------------------------------------------------------------|
| Jonas Fejes  | IVL Swedish Environmental Research<br>Institute, Business and Development<br>Area Water IVL Oil Spill, Box 21 060,<br>Stockholm, SE-100 31, Sweden |

Tel: +46 8 598 563 00
Fax: +46 8 598 56 390
E-mail: Jonas.fejes@ivl.se

Andrei Kozeltsev

Ministry of Natural Resources
4/6 B. Gruzinskaya st., Moscow,
123995, D-242, GSP-5m, Russia
Tel: (7-095) 254 0229
Fax: (7-095) 943 0013
E-mail: Kozeltsev@mnr.gov.ru

Vladimir Krivilev

Academy for Geopolitical Problems, 5-
ay Magistrinya Str. 10/2, off. 21,
Moscow, 123007, Russia
E-mail: vkrivilev@sops.ru

Sergei Ovsienko

Head of Laboratory Russian State
Oceanography Institute, Kropotkinsky
per. 6, Moscow, 119034, Russia
Tel: +7 495 246 7288
E-mail: snovs@orc.ru

Dr. Andrey D Samatov

Sakhalin Scientific Research Institute of
Fisheries and Oceanography, Off. 307,
196, Komsomolskaya St.,
Yuzhno-Sakhalinsk 693023, Russia
Tel: +7(4242) 45 67 67
Fax: +7(4242) 45 67 78
E-mail: samatov@sakhniro.ru

## Other Speakers from Partner Countries

Darko Blinkov

State Environmental Inspector
IMPEL (EU-Implementation on
Environmental Law) and ECENA
(Enforcement and Compliance
Environmental Network for
Accession) Deputy Coordinator
for RM
Ministry of Environment and,

Physical planning, State
Environmental Inspectorate
Drezdenska 52, 1000 Skopje,
The Former Yugoslav Republic of
Macedonia
Tel: +389 2 3066 930 ext. 271
Fax: +389 2 3066 931
E-mail:
D.Blinkov@moepp.gov.mk

Kosta Trajkovski

Ministry of Environment and
Physical Planning "Drezdenska"
52, Skopje, 1000
The Former Yugoslav Republic of
Macedonia
Tel: (389-2) 3066 930
Fax: (389-2) 3066 931
E-mail:
K.Trajkovski@moepp.gov.mk